Introduction to Noncommutative Algebra

Introduction to Noncommutative Algebra

Editor

Linsen Chou

scitus
academics

Introduction to Noncommutative Algebra

Edited by **Linsen Chou**

Printed in 2017

ISBN: 978-1-68117-188-3

Library of Congress Control Number: 2015949133

SCITUS Academics LLC,
616, Corporate Way, Suite 2, 4766,
Valley Cottage, NY 10989

www.scitusacademics.com

Preface

A noncommutative algebra is an associative algebra in which the multiplication is not commutative, that is, for which xy does not always equal yx; or more generally an algebraic structure in which one of the principal binary operations is not commutative; one also allows additional structures, e.g. topology or norm, to be possibly carried by the noncommutative algebra of functions.

The main motivation is to extend the commutative duality between spaces and functions to the noncommutative setting. In mathematics, spaces, which are geometric in nature, can be related to numerical functions on them. In general, such functions will form a commutative ring. For instance, one may take the ring C(X) of continuous complex-valued functions on a topological space X. In many cases, we can recover X from C(X), and therefore it makes some sense to say that X has commutative topology.

The dream of noncommutative geometry is to generalize this duality to the duality between noncommutative algebras, or sheaves of noncommutative algebras, or sheaf-like noncommutative algebraic or operator-algebraic structures and geometric entities of certain kind, and interact between the algebraic and geometric description of those via this duality.

Regarding that the commutative rings correspond to usual affine schemes, and commutative C*-algebras to usual topological spaces, the extension to noncommutative rings and algebras requires non-trivial generalization of topological spaces, as "non-commutative spaces". This book provides an elementary introduction to noncommutative rings and algebras,

Table of Contents

Free Quasi-symmetric Functions and Descent Algebras for Wreath Products, and Noncommutative Multi-symmetric Functions

1

Jean-Christophe Novelli and
Jean-Yves Thibon
Université Paris-Est, Institut Gaspard Monge,
5 Boulevard Descartes, Champs-sur-Marne,
77454 Marne-la-Vallée cedex 2, France

ABSTRACT

We introduce analogs of the Hopf algebra of free quasi-symmetric functions with bases labeled by colored permutations. When the color set is a semigroup, an internal product can be introduced. This leads to the construction of generalized descent algebras associated with wreath products $\Gamma \wr \mathfrak{S}_n$ and to the corresponding generalizations of quasi-symmetric functions. The associated Hopf algebras appear as natural analogs of McMahon's multisymmetric functions. As a consequence, we obtain an internal product on ordinary multi-symmetric functions. We extend these constructions to Hopf algebras of colored parking functions, colored non-crossing partitions and parking functions of type B.

INTRODUCTION

The Hopf algebra of Free Quasi-Symmetric Functions FQSym [25] and [4] is an algebra of noncommutative polynomials associated with the sequence $(\mathfrak{S}_n)_{n\geq 0}$ of all symmetric groups. It is connected by Hopf homomorphisms to several other important algebras associated with the same sequence of groups: free symmetric functions (or coplactic algebra) FQSym [35], [4] and [2], non-commutative symmetric functions (or descent algebras) Sym [7], Quasi-Symmetric functions QSym [9], Symmetric functions Sym, and also, Planar binary trees PBT [22] and [12], Matrix quasi-symmetric functions MQSym [4] and [11], Parking functions PQSym [18] and [29], and so on.

Most of these Hopf algebras are endowed with an internal product, generalizing the one of ordinary symmetric functions. The basic example is provided by noncommutative symmetric functions, whose homogeneous components can be identified with the Solomon descent algebras of symmetric groups [7].

Symmetric groups are the Coxeter groups of type A, and there are descent algebras for other types as well. However, the direct sums of the descent algebras of types B or D are not Hopf algebras in any natural way. But there are Hopf algebras associated with wreath products $Z_\ell \wr \mathfrak{S}_n$, the Mantaci–Reutenauer algebras [26], which admit internal products, and contain the Solomon algebras of type B for $\ell = 2$.

From the point of view of symmetric functions, $MR^{(\ell)}$, the Mantaci–Reutenauer algebra of level ℓ is the free product of ℓ copies of Sym. It is therefore the natural noncommutative analog of $(Sym)^{\otimes \ell} \simeq Sym(X_0;\cdots;X_{\ell-1})$, the algebra of symmetric functions in ℓ independent alphabets, which is also the Grothendieck ring of the tower of algebras $(C[Z_\ell \wr \mathfrak{S}_n])$. And indeed, it has been shown in [14] that $MR^{(\ell)}$ was the Grothendieck ring of projective modules over the 0-Ariki–Koike–Shoji algebras, a degeneracy of the Hecke algebras associated with $Z_\ell \wr \mathfrak{S}_n$.

However, with ℓ independent alphabets, one can build a larger Hopf algebra. In the commutative case, it is the algebra of multi-symmetric functions, first introduced by McMahon [27], and briefly investigated by Gessel from a modern point of view in [10]. It is defined as follows. Setting $X_i = \{x_{i,j} | j = 1, \ldots, n\}$, the multi-symmetric polynomials are the invariants of $\mathfrak{S}_n$ in $C[X_0, \ldots, X_{\ell-1}]$ for the diagonal action (by the automorphisms $\sigma(x_{i,j}) = x_{i,\sigma(j)}$). This is an algebra, which, as usual, acquires a Hopf algebra structure in the limit $n \to \infty$.

In the following, we will start with a level ℓ analogue of FQSym, whose bases are labeled by ℓ-colored permutations. Imitating the embedding of Sym in FQSym, we obtain a Hopf subalgebra of level ℓ called $Sym^{(\ell)}$, which is a natural noncommutative analog of McMahon's algebra of multi-symmetric functions, and turns out to be dual to Poirier's quasi-symmetric functions. Its homogenous components can be endowed with an internal product, thus providing an analog of Solomon's descent algebras for wreath products, bigger than the Mantaci–Reutenauer algebras, and in which most useful properties such as the splitting formula remain valid. On the commutative image, this yields an internal product on multi-symmetric functions.

The Mantaci–Reutenauer descent algebra $MR^{(\ell)}$ arises as a natural Hopf subalgebra of $Sym^{(\ell)}$ and its dual is computed in a straightforward way by means of an appropriate Cauchy formula.

Finally, we introduce a Hopf algebra of colored parking functions, and use it to define Hopf algebras structures on parking functions and noncrossing partitions of type B.

The main results of this paper have been announced in the draft [30]. Since then, some of these results, in particular the construction of $Sym^{(\ell)}$, have been used by Baumann and Hohlweg [1], whose paper provides detailed proofs. Hence, we shall only include the proofs which cannot be found in their paper. In particular, we propose an alternative approach to the internal product, which is introduced by a duality argument, and derive its main properties from those of the dual coproduct.

BACKGROUND AND NOTATIONS

We first explain how to adapt the classical definitions and operations to the ℓ-colored case.

Colored Alphabets

We shall start with an ℓ -colored alphabet

$$\mathbf{A} = A^0 \sqcup A^1 \sqcup \cdots \sqcup A^{\ell-1}, \tag{1}$$

such that all A^i are of the same cardinality N, which will be assumed to be infinite in the sequel. Let C be the alphabet $\{ C_0, \cdots, C_{\ell-1} \}$ and A be the auxiliary ordered alphabet $\{1, 2, \ldots\}$ (the letter C stands for colors and A for alphabet) so that A can be identified with the Cartesian product $A \times C$:

$$\mathbf{A} \simeq A \times C = \{(a, c), a \in A, c \in C\}. \tag{2}$$

A colored letter (i, c) will be denoted in bold type i. Words over A are called colored words. Given two colored words, their concatenation is obtained by concatenating separately the elements coming from A and from C. We will sometimes allow $\ell = \infty$.

Colored Standardization

Let w be a word over A, represented as $(\mathcal{V}, \mathcal{U})$ with $\mathcal{V} \in A^*$ and $\mathcal{U} \in C^*$. Then the colored standardized word Std($\mathcal{W}$) of $\mathcal{W}$ is

$$\mathbf{Std}(w) := (\mathrm{Std}(v), u), \tag{3}$$

where Std($\mathcal{V}$) is the usual standardization on words.

Recall that the standardization process sends a word $\mathcal{V}$ of length n to a permutation Std $(\mathcal{V}) \in \mathfrak{S}_n$, called the standardized of v, defined as the permutation obtained by iteratively scanning v from left to right, and labeling 1,2,... the occurrences of its smallest letter, then numbering the occurrences of the next one, and so on. Alternatively, Std $(\mathcal{V})$ is the permutation having the same inversions as $\mathcal{V}$.

For example, Std(abcadbdaa) = 157286934.

$$\begin{matrix} a & b & c & a & d & b & d & a & a \\ a_1 & b_5 & c_7 & a_2 & d_8 & b_6 & d_9 & a_3 & a_4 \\ 1 & 5 & 7 & 2 & 8 & 6 & 9 & 3 & 4 \end{matrix} \quad (4)$$

so that

$$\mathbf{Std}(abcadbdaa, 144120100) = (157286934, 144120100). \quad (5)$$

Colored Shifted Operations

For an element $\mathcal{V} = (\mathcal{V}_1, \mathcal{V}_2, \ldots)$ of A and an integer k, denote by $\mathcal{V}[k]$ the shifted word $(\mathcal{V}_1 + k) \cdots (\mathcal{V}_n + k)$, e.g. $312[4] = 756$. Given a colored word $\alpha = (\alpha, u), \alpha[k] = (\alpha[k], u)$, we set.

The shifted concatenation of two words v and v′ is defined by

$$v \bullet v' := v \cdot v'[k] \quad (6)$$

where k is the length of $\mathcal{V}$.

The shifted concatenation of two colored words $(v, c), (v', c')$ is defined by

$$(v, c) \bullet (v', c') := (v \cdot v'[k], c \cdot c') \quad (7)$$

For example,

$$(13241, 00322) \bullet (12, 23) = (1324167, 0032223). \quad (8)$$

Finally, recall that the shuffle product of two words au and b$\mathcal{V}$ is defined by

$$au \sqcup\!\sqcup bv = a(u \sqcup\!\sqcup bv) + b(au \sqcup\!\sqcup v), \quad (9)$$

where a and b are letters and u and v are words, with the initial conditions

$$u \sqcup\!\sqcup \epsilon = \epsilon \sqcup\!\sqcup u = u, \tag{10}$$

being the empty word.

This extends to the colored case, considering colored words as concatenations of colored letters.

The shifted shuffle product is

$$u \Cup v := u \sqcup\!\sqcup v[k], \tag{11}$$

where k is the size of u.

FREE QUASI-SYMMETRIC FUNCTIONS OF LEVEL ℓ

FQSym$^{(\ell)}$ and FQSym$^{(r)}$

A colored permutation is a pair (σ, u), with $\sigma \in \mathfrak{S}_n$ and $u \in C^n$, the integer n being the size of this permutation.

Definition 3.1

The dual free colored quasi-ribbon $G_{\sigma,u}$ labeled by a colored permutation (σ, u) of size n is the noncommutative polynomial

$$\mathbf{G}_{\sigma,u} := \sum_{w \in A^n; \mathbf{Std}(w) = (\sigma,u)} w \quad \in Z\langle \mathbf{A} \rangle. \tag{12}$$

Recall that the convolution of two permutations σ and μ is the set $\sigma * \mu$ (identified with the formal sum of its elements) of permutations τ such that the standardized word of the $|\sigma|$ first letters of τ is σ and the standardized word of the remaining letters of τ is μ (see [38]). We then have:

Theorem 3.2

Let (σ',u') and (σ'',u'') be colored permutations. Then

$$\mathbf{G}_{\sigma',u'}\mathbf{G}_{\sigma'',u''} = \sum_{\sigma\in\sigma'*\sigma''} \mathbf{G}_{\sigma,u'\cdot u''}. \tag{13}$$

Therefore, the dual free colored quasi-ribbons span a Z-subalgebra $\mathbf{FQSym}^{(\ell)}$ of the free associative algebra.

Proof

This is immediate from the product of the usual free quasi-symmetric functions G_σ:

$$\mathbf{G}_{\sigma'}\mathbf{G}_{\sigma''} = \sum_{\sigma\in\sigma'*\sigma''} \mathbf{G}_{\sigma}. \tag{14}$$

Note that all colored permutations indexing a product of G have given colors at the same places. For example,

$$\mathbf{G}_{(21,41)}\mathbf{G}_{(12,31)} = \mathbf{G}_{(2134,4131)} + \mathbf{G}_{(3124,4131)} + \mathbf{G}_{(4123,4131)} + \mathbf{G}_{(3214,4131)} + \mathbf{G}_{(4213,4131)} + \mathbf{G}_{(4312,4131)}. \tag{15}$$

One can define a coproduct by the usual trick of sums of alphabets: observe that we only need a total order on A to define the colored standardization, so that taking two isomorphic copies A' and A'' of A, we define $A' + A''$ as $(A''+A'')\times C$, where $A'+A''$ denotes the ordinal sum, that is, the disjoint union $A'\sqcup A''$ endowed with an order which extends those of A' and A'', and such that $a'<a''$ for all $a'\in A'$ and $a''\in A''$. Note that this operation is not commutative. Assuming furthermore that A' and A'' commute, we identify $f(A')$ and $g(A'')$ with $f\otimes g$ and define a coproduct by:

$$\Delta f(A) = f(A' + A''). \tag{16}$$

By construction, this is an algebra morphism from $\mathbf{FQSym}^{(\ell)}$ to $\mathbf{FQSym}^{(\ell)} \otimes \mathbf{FQSym}^{(\ell)}$ so that

Theorem3.3

$\mathrm{FQSym}^{(\ell)}$ is a graded connected bialgebra. Hence, it is a Hopf algebra. The coproduct is given by

$$\Delta \mathbf{G}_{\sigma,u} := \sum_{\substack{(\sigma',\sigma'',u',u'') \\ (\sigma,u)\in(\sigma',u')\Cup(\sigma'',u'')}} \mathbf{G}_{(\sigma',u')} \otimes \mathbf{G}_{(\sigma'',u'')}. \tag{17}$$

For example,

$$\Delta \mathbf{G}_{3142,2412} = 1 \otimes \mathbf{G}_{3142,2412} + \mathbf{G}_{1,4} \otimes \mathbf{G}_{231,212} + \mathbf{G}_{12,42} \otimes \mathbf{G}_{12,21} + \mathbf{G}_{312,242} \otimes \mathbf{G}_{1,1} + \mathbf{G}_{3142,2412} \otimes 1. \tag{18}$$

Proof

This is again immediate from the co product of the usual free quasi-symmetric functions G_σ:

$$\Delta \mathbf{G}_\sigma = \sum_{\sigma \in \sigma' \Cup \sigma''} \mathbf{G}_{\sigma'} \otimes \mathbf{G}_{\sigma''}. \tag{19}$$

Taking colored letters instead of usual letters does not change the expression of $G(A' + A'')$, the colors following the letter they are attached to, as in the definition of the shifted shuffle in the colored case.

Duality in $\mathbf{FQSym}^{(\ell)}$

Definition 3.4

The free ℓ-quasi-ribbon $F_{\sigma,u}$ labeled by a colored permutation (σ, u) is the no commutative polynomial

$$\mathbf{F}_{\sigma,u} := \mathbf{G}_{\sigma^{-1},u\cdot\sigma^{-1}}, \tag{20}$$

where the action of a permutation on the right of a word permutes the positions of the letters of the word.

For example,

$$\mathbf{F}_{3142,2142} = \mathbf{G}_{2413,1422}. \tag{21}$$

The product and coproduct of the $F_{\sigma,u}$ can be easily described in terms of shifted shuffle and deconcatenation of colored permutations.

Theorem 3.5

Let σ' and σ'' be two colored permutations. Then

$$\mathbf{F}_{\sigma'}\mathbf{F}_{\sigma''} = \sum_{\sigma\in\sigma'\Cup\sigma''} \mathbf{F}_{\sigma}, \tag{22}$$

and

$$\Delta\mathbf{F}_{\sigma} = \sum_{\substack{w',w'' \\ \sigma=w'.w''}} \mathbf{F}_{\mathbf{Std}(w')} \otimes \mathbf{F}_{\mathbf{Std}(w'')}. \tag{23}$$

Proof

Without colors, these formulas are the usual product or coproduct formulas of the F in FQSym. With colors, one just has to observe that colors follow the letter to which they are attached.

Note that all colored permutations indexing a product of F have given colors associated with the same values, which is consistent with the corresponding remark on the G, since places and values are exchanged when taking the inverse of a permutation.

For example, compare the following with Eq. (15):

$$\mathbf{F}_{(21,14)}\mathbf{F}_{(12,31)} = \mathbf{F}_{(2134,1431)} + \mathbf{F}_{(2314,1341)} + \mathbf{F}_{(2341,1314)} + \mathbf{F}_{(3214,3141)} + \mathbf{F}_{(3241,3114)} + \mathbf{F}_{(3421,3114)}. \tag{24}$$

Here is an example of coproduct on the F basis:

$$\begin{aligned}\Delta\mathbf{F}_{(23514,14212)} = \; & 1 \otimes \mathbf{F}_{(23514,14212)} + \mathbf{F}_{(1,1)} \otimes \mathbf{F}_{(2413,4212)} + \mathbf{F}_{(12,14)} \otimes \mathbf{F}_{(312,212)} \\ & + \mathbf{F}_{(123,142)} \otimes \mathbf{F}_{(12,12)} + \mathbf{F}_{(2341,1421)} \otimes \mathbf{F}_{(1,2)} + \mathbf{F}_{(23514,14212)} \otimes 1\end{aligned} \tag{25}$$

Let us define a scalar product on $\mathrm{FQSym}^{(\ell)}$ by

$$\langle \mathbf{F}_{\sigma,u}, \mathbf{G}_{\sigma',u'} \rangle := \delta_{\sigma,\sigma'}\delta_{u,u'}, \tag{26}$$

where δ is the Kronecker symbol.

Theorem 3.6

For any U, V, W ∈ $\mathrm{FQSym}^{(\ell)}$,

$$\langle \Delta U, V \otimes W \rangle = \langle U, VW \rangle, \tag{27}$$

so that $\mathrm{FQSym}^{(\ell)}$ is self-dual: the map $F_{\sigma,u}$ $F^*_{\sigma,u}$ is a Hopf algebra isomorphism from $\mathrm{FQSym}^{(\ell)}$ to its graded dual.

Proof

Formula (27) is straightforward from Theorem 3.5. Then, in particular,

$$\langle \Delta\mathbf{F}_{\sigma,u}, \mathbf{G}_{\alpha,v} \otimes \mathbf{G}_{\beta,w} \rangle = \langle \mathbf{F}_{\sigma,u}, \mathbf{G}_{\alpha,v}\mathbf{G}_{\beta,w} \rangle, \tag{28}$$

so that G can be identified with the dual basis F* of F.

Note 3.7: Let ϕ be any bijection from C to C, extended to words by concatenation. Then if one defines the free ℓ-quasi-ribbon as

$$\mathbf{F}_{\sigma,u} := \mathbf{G}_{\sigma^{-1},\phi(u)\cdot\sigma^{-1}}, \tag{29}$$

the previous theorems remain valid since one only permutes the labels of the basis ($F_{\sigma, u}$). Moreover, if C has a semigroup structure, the colored permutations $(\sigma, u) \in \mathfrak{S}_n \times C^n$ can be interpreted as elements of the semi-direct product $H_n := \mathfrak{S}_n \ltimes C^n$ with multiplication rule

$$(\sigma; c_1, \ldots, c_n) \cdot (\tau; d_1, \ldots, d_n) := (\sigma\tau; c_{\tau(1)}d_1, \ldots, c_{\tau(n)}d_n). \tag{30}$$

If furthermore C is a group, one can choose $\phi(\gamma) := \gamma^{-1}$ and define the scalar product as before, so that the adjoint basis of the (G_h) becomes $F_h := G_{h^{-1}}$. In the sequel, we will be mainly interested in the cases $C := Z/\ell Z$, and we will indeed make that choice for ϕ whenever C is a group.

Algebraic Structure of FQSym(ℓ)

Recall that a permutation σ of size n is connected [26] and [4] if, for any $i<n$, the set $\{\sigma_1, \ldots, \sigma_i\}$ is different from $\{1, \ldots, i\}$.

We denote by $\mathcal{C}$ the set of connected permutations, and by $C_n := |\mathcal{C}_n|$ the number of such permutations in $\mathfrak{S}_n$. For later reference, we recall that $\mathcal{C}_n$ is Sequence A003319 of [40], whose generating series is

$$c(t) := \sum_{n\geq 1} c_n t^n = 1 - \left(\sum_{n\geq 0} n! t^n\right)^{-1} = t + t^2 + 3t^3 + 13t^4 + 71t^5 + 461t^6 + O(t^7). \tag{31}$$

Let the connected colored permutations be the (σ, u) with σ connected and u arbitrary. Their generating series is given by $c(\ell t)$.

From [4], we get immediately

Proposition 3.8

FQSym(ℓ) is free over the set $F_{\sigma, u}$ (or $G_{\sigma, u}$), where (σ,u) is connected.

For example, the generating series of the algebraic generators of $\mathrm{FQSym}^{(2)}$ is

$$2t + 4t^2 + 24t^3 + 208t^4 + 2272t^5 + 29504t^6 + 441216t^7 + \cdots \qquad (32)$$

This is sequence A141307 of [40].

Dendriform Structure of FQSym(ℓ)

Foissy introduced the notion of bidendriform bialgebras [6], generalizing the notion of dendriform algebras (cf. [21]) and proved some conjectures about FQSym, presented in [4]. We shall adapt this technology to the colored case. We shall not recall all the theory, since complete details can be found in [6].

Recall that the generators of FQSym(ℓ) as a dendriform algebra are called totally primitive elements [6]and that their generating series is given by

$$\mathrm{TP} := \frac{\mathrm{PQ} - 1}{\mathrm{PQ}^2}, \qquad (33)$$

where PQ is the generating series of FQSym(ℓ).

Theorem 3.9
The algebra FQSym(ℓ) has a structure of bidendriform bialgebra [6], hence is free as an associative algebra and as a dendriform algebra, cofree, self-dual, and its primitive Lie algebra is free.
Moreover, the totally primitive elements of FQSym(ℓ) are the totally primitive elements of FQSym with any coloring.

Proof

It has been done by Foissy in [6] in the case of FQSym. But since the dendriform and codendriform structure do not involve the color alphabet C, the property is true in this case as well.

Similarly, colors do not play any role in determining if a given element is (totally) primitive or not.

For example, the dendriform generators of FQSym have as degree generating series

$$\sum_i \mathrm{dg}_i t^i = t + t^3 + 6t^4 + 39t^5 + 284t^6 + 2305t^7 + \cdots \tag{34}$$

(cf. [40, A122827]), so that the dendriform generators of $\mathrm{FQSym}^{(2)}$ have as degree generating series $2^i \mathrm{dg}_i$.

$$2t + 8t^3 + 96t^4 + 1248t^5 + 18176t^6 + 295\,040t^7 + \cdots . \tag{35}$$

Note that there cannot be any relation, even dendriform relations, among the elements F1,c where c∈C, so that FQSym(ℓ) contains the free dendriform algebra PBT(ℓ) on ℓ generators. This interpretation has been exploited in [33].

Primitive Elements of FQSym(ℓ)

Let $\mathcal{L}^{(\ell)}$ be the primitive Lie algebra of $\mathrm{FQSym}^{(\ell)}$. Since Δ is not cocommutative, $\mathrm{FQSym}^{(\ell)}$ cannot be the universal enveloping algebra of $\mathcal{L}^{(\ell)}$. But since it is cofree, it is, according to [23], the universal enveloping dipterous algebra of its primitive part $\mathcal{L}^{(\ell)}$.

Let $G^{\sigma,u}$ be the multiplicative basis defined by $G^{\sigma,u} = G_{\sigma 1,u1}\ldots\ldots F_{\sigma r,ur}$ where $(\sigma,u)=(\sigma_1,u_1)\bullet\cdots\bullet(\sigma_r,u_r)$ is the unique maximal factorization of $(\sigma,u)\in \mathfrak{S}_n\times C^n$ into connected colored permutations.

Proposition 3.10

Let $G_{\sigma,u}$ be the adjoint basis of $G^{\sigma,u}$. Then, the family $(V_{\alpha,u})_{\alpha\in C}$ is a basis of $\mathcal{L}^{(\ell)}$. In particular, we have $\dim\mathcal{L}_n^{(\ell)} = \ell^n c_n$ Moreover, $\mathcal{L}^{(\ell)}$ is free.

Proof

The first part of the statement follows from [4, Prop. 3.6]. The second part comes from the fact that FQSym$^{(\ell)}$ is bidendriform (Theorem 3.9), since, in any bidendriform bialgebra, the primitive Lie algebra is free.

For example, since $\mathcal{L}^{(\ell)}$ is free, the generating series by degree of its generators is (with $\ell = 2$):

$$1-\prod_{n\geq 1}(1-t^n)^{\ell^n c_n} = 1-(1-t)^2(1-t^2)^4(1-t^3)^{24}\cdots$$
$$= 2t+3t^2+16t^3+158t^4+1796t^5+24\,250t^6+372\,656t^7+\cdots \tag{36}$$

since the Hilbert series of its universal enveloping algebra (the domain of cocommutativity of FQSym$^{(\ell)}$) is, again with $\ell = 2$ [40, A141309],

$$\prod_{n\geq 1}(1-t^n)^{-\ell^n c_n} = 1+2t+7t^2+36t^3+283t^4+2898t^5+36\,169t^6+524\,976t^7+\cdots. \tag{37}$$

Indeed, the dimension of $\mathcal{L}^{(\ell)}$ in degree n being $\ell^n c_n$, the Hilbert series of $U(\mathcal{L}^{(\ell)})$ coincides with that of its symmetric algebra $S(\mathcal{L}^{(\ell)})$ by the PBW theorem. ℓ

Internal Product of FQSym(ℓ)

When C is a semigroup, an internal product can be defined on FQSym$^{(\ell)}$ by

$$\mathbf{F}_{\sigma,u} * \mathbf{F}_{\tau,v} = \mathbf{F}_{\mu,w} \tag{38}$$

where (μ, w) is the product $(\sigma, u)\cdot(\tau, v)$ in the wreath product, defined by formula (30), that is

$$\mathbf{F}_{(\sigma,u)} * \mathbf{F}_{(\tau,v)} = \mathbf{F}_{(\sigma\tau,(u\tau)\cdot v)}, \tag{39}$$

where $u\tau$ is the word $u_{\tau_1}\cdots u_{\tau_n}$ and $u\cdot v$ is the componentwise product defined by $(u_1v_1,\ldots,u_nv_n)$.

For example, if the color group is Z

$$\mathbf{F}_{(1324,1011)} * \mathbf{F}_{(2413,3200)} = \mathbf{F}_{(3412,3311)} \tag{40}$$

$$\mathbf{F}_{(165324,102011)} * \mathbf{F}_{(625413,322011)} = \mathbf{F}_{(462315,423023)}. \tag{41}$$

This can be reduced to any $Z/\ell Z$, e.g. , with $\ell = 3$,

$$\mathbf{F}_{(165324,102011)} * \mathbf{F}_{(625413,022011)} = \mathbf{F}_{(462315,120020)}. \tag{42}$$

In the G basis, one has

$$\mathbf{G}_{(\sigma,u)} * \mathbf{G}_{(\tau,v)} = \mathbf{G}_{(\tau\sigma,u.(v\sigma))}, \tag{43}$$

NONCOMMUTATIVE SYMMETRIC FUNCTIONS OF LEVEL ℓ

ℓ-Partite Numbers

Following McMahon [27], we define an ℓ-partite number $n = (n_1, \cdots, n_\ell)$ as a column vector in N^ℓ, and avector composition of n of weight $|n| := \sum_i n_i$ and length m as a $\ell \times m$ matrix $\mathbf{I}$ of nonnegative integers, with row sums vector $\mathbf{n}$ and no zero column.

For example,

$$\mathbf{I} = \begin{pmatrix} 1 & 0 & 2 & 1 \\ 0 & 3 & 1 & 1 \\ 4 & 2 & 1 & 3 \end{pmatrix} \tag{44}$$

is a vector composition (or a 3-composition, for short) of the 3-partite number $\begin{pmatrix} 4 \\ 5 \\ 10 \end{pmatrix}$ of weight 19 and length 4.

For each $\mathbf{n} \in \mathbb{N}^{\ell}$ of weight $|\mathbf{n}| = n$, we define a level ℓ complete homogeneous noncommutative symmetric function as

$$S_{\mathbf{n}} := \sum_{u;\, m_i(u)=n_i} \mathbf{G}_{1\cdots n,u}, \tag{45}$$

where $m_i(u)$ denotes the number of occurrences of i in u. It is the sum of all possible colorings of the identity permutation with n_i occurrences of color i for each i.

The Hopf Algebra $\mathrm{Sym}^{(\ell)}$

Let $\mathrm{Sym}^{(\ell)}$ be the subalgebra of $\mathrm{FQSym}^{(\ell)}$ generated by the $S_{\mathbf{n}}$ (with the convention $S_0 = 1$).

Theorem 4.1

$\mathrm{Sym}^{(\ell)}$ is free over the set $\{S_{\mathbf{n}}, |\mathbf{n}| > 0\}$, so that a linear basis is given by

$$S^{\mathbf{I}} = S_{\mathbf{i}_1} \cdots S_{\mathbf{i}_m}, \tag{46}$$

where $\mathbf{i}_1, \cdots, \mathbf{i}_m$ are the columns of I.

Moreover, $\mathrm{Sym}^{(\ell)}$ is a Hopf subalgebra of $\mathrm{FQSym}^{(\ell)}$ and the coproduct of the generators is given by

$$\Delta S_{\mathbf{n}} = \sum_{\mathbf{i}+\mathbf{j}=\mathbf{n}} S_{\mathbf{i}} \otimes S_{\mathbf{j}}, \tag{47}$$

where the sum i+j is taken in the space NI. In particular, $\mathrm{Sym}^{(\ell)}$ is cocommutative.

Proof: Consider a linear relation between the SI. Since the same relation would hold without the colors, this implies a relation within Sym. But since Sym is free, that means that the initial linear relation splits into linear relations involving only vector compositions having same column sums. But all these linear relations are trivial thanks to Formula (45): all such SI have no word in common in the realization of $\mathrm{Sym}^{(\ell)}$ in the free algebra over colored letters. Indeed, given an SI , one uniquely rebuilds the parts $i_1, i_2 \cdots, i_m$ as the number of occurrences of each color of any word in SI respectively in the first $|i_1|$ letters, the next $|i_2|$ letters, and so on.

For example,

$$\Delta S^{\begin{pmatrix}1\\0\\2\end{pmatrix}} = S^{\begin{pmatrix}1\\0\\2\end{pmatrix}} \otimes S^{\begin{pmatrix}0\\0\\0\end{pmatrix}} + S^{\begin{pmatrix}0\\0\\2\end{pmatrix}} \otimes S^{\begin{pmatrix}1\\0\\0\end{pmatrix}} + S^{\begin{pmatrix}1\\0\\1\end{pmatrix}} \otimes S^{\begin{pmatrix}0\\0\\1\end{pmatrix}} + S^{\begin{pmatrix}0\\0\\1\end{pmatrix}} \otimes S^{\begin{pmatrix}1\\0\\1\end{pmatrix}} + S^{\begin{pmatrix}1\\0\\0\end{pmatrix}} \otimes S^{\begin{pmatrix}0\\0\\2\end{pmatrix}} + S^{\begin{pmatrix}0\\0\\0\end{pmatrix}} \otimes S^{\begin{pmatrix}1\\0\\2\end{pmatrix}} \quad (48)$$

Algebraic Structure of $\mathrm{Sym}^{(\ell)}$

The number of generators of $\mathrm{Sym}^{(\ell)}$ of degree n is given by the number of -partite numbers of total sum n. So its generating series is

$$(1-t)^{-\ell} - 1 = \sum_{n \geq 1} \binom{\ell + n - 1}{n} t^n \quad (49)$$

Hence, the Hilbert series of $\mathrm{Sym}^{(\ell)}$ is

$$S_\ell(t) := \sum_n \dim \mathbf{Sym}_n^{(\ell)} t^n = \frac{(1-t)^\ell}{2(1-t)^\ell - 1} \quad (50)$$

For example, with $\ell = 2$, one has

$$S_2(t) := 1 + 2t + 7t^2 + 24t^3 + 82t^4 + 280t^5 + 956t^6 + 3264t^7 + \cdots \quad (51)$$

which is Sequence A003480 of [40].

For general ℓ, it is well-known in the combinatorial folklore (and easy to prove by means of generating series expansions) that the coefficients of this series satisfy $n \geq 1$,

$$n!\ \dim(\mathbf{Sym}_n^{(\ell)}) = \sum_{k=1}^{n} S(n,k) p_k \ell^k \quad (52)$$

where $S(n,k)$ is the sequence of absolute values of Stirling numbers of the first kind (sequence A130534 of [40]) and p_k is the sequence of ordered Bell numbers (also known as packed words or preferential arrangements, Sequence A000670 of [40]).

We shall denote by $G(\ell)$ the set of nonzero ℓ-partite numbers.

Primitive Elements of $\mathrm{Sym}^{(\ell)}$

$\mathrm{Sym}^{(\ell)}$ **being a graded connected cocommutative Hopf algebra, it follows from the Cartier–Milnor–Moore theorem that it is the universal enveloping algebra of $L(^{\ell})$:**

$$\mathbf{Sym}^{(\ell)} = U(L^{(\ell)}) \quad (53)$$

where $L^{(\ell)}$ is the Lie algebra of its primitive elements. Let us now prove.

Theorem 4.2

As a graded Lie algebra, the primitive Lie algebra L($^\ell$) of $\mathrm{Sym}^{(\ell)}$ is free over a set indexed by G(ℓ).

Proof: For each generator Sn of $\mathrm{Sym}^{(\ell)}$, there exists a primitive element of the form

$$\Phi_{\mathbf{n}} = S_{\mathbf{n}} + \sum_{l(I)\geq 2} c_I S^{\mathbf{I}} \tag{54}$$

Indeed, one can take

$$\Phi_{\mathbf{n}} := \pi_1(S_{\mathbf{n}}) \tag{55}$$

where π1 the Eulerian idempotent, that is, the endomorphism of $\mathrm{Sym}^{(\ell)}$ defined by $\pi_1 = \log^*(\mathrm{Id})$ (see [38, 58]). The coproduct formula (47) ensures that Φ_n is of the form (54). Such elements are obviously algebraically independent, hence generate a free Lie algebra L′ contained in L($^\ell$). But Formula (53) shows that the Hilbert series of L(ℓ) must be the same as that of L′, so that L′=L(ℓ).

The generating series of the dimensions of $L^{(\ell)}$ is, according to (49)

$$D(t) := \sum_{n\geq 1} d_n(\ell) t^n, \tag{56}$$

where

$$\prod_{n\geq 1} \left(\frac{1}{1-t^n}\right)^{d_n(\ell)} = S_\ell(t) \tag{57}$$

For example, with $\ell=2$, the generating series of the dimensions of $L^{(\ell)}$ is

$$2t+4t^2+12t^3+31t^4+92t^5+256t^6+772t^7+\cdots \tag{58}$$

with $\ell=3$, one finds

$$3t+9t^2+36t^3+132t^4+534t^5+2140t^6+8982t^7+\cdots. \tag{59}$$

Duality in $Sym^{(\ell)}$

Recall that the underlying colored alphabet A can be seen as $A^0 \sqcup \cdots \sqcup A^{\ell-1}$, with $A^i=\{a_j^{(i)}, j\geq 1\}$. Let $X=(x^{(0)},\ldots,x^{(\ell-1)})$, where the $x^{(i)}$ are ℓ commuting variables. In terms of A, the generating function of the complete functions can be written as

$$\sigma_{\mathbf{x}}(A)=\prod_{i\geq 1}^{\rightarrow}\left(1-\sum_{0\leq j\leq \ell-1} x^{(j)}a_i^{(j)}\right)^{-1}=\sum_{\mathbf{n}} S_{\mathbf{n}}(A)\mathbf{x}^{\mathbf{n}} \tag{60}$$

where

$$\mathbf{x}^{\mathbf{n}}=(x^{(0)})^{n_0}\cdots(x^{(\ell-1)})^{n_{\ell-1}}.$$

This realization gives rise to a Cauchy formula (see [19] for the l=1 case), which in turn allows one to identify the dual of $Sym^{(\ell)}$ with an algebra introduced by Poirier in [34]. It is detailed in the following section.

Note that $Sym^{(\ell)}$ is the natural noncommutative analog of McMahon's algebra of multisymmetric functions [27] and [10].

QUASI-SYMMETRIC FUNCTIONS OF LEVEL ℓ

Cauchy Formula of Level ℓ

Let $\mathbb{X} = X^0 \sqcup \cdots \sqcup X^{\ell-1}$, where $X^i = \left\{x_j^{(i)}, j \geq 1\right\}$, be an ℓ-colored alphabet of commutative variables, also commuting with A. Imitating the level 1 case (see [4]), we define the Cauchy kernel

$$K(\mathbb{X}, \boldsymbol{A}) = \prod_{j\geq 1}^{\rightarrow} \sigma_{\left(x_j^{(0)},\ldots,x_j^{(\ell-1)}\right)}(\boldsymbol{A}) \qquad (61)$$

Expanding on the basis $S^{\mathbf{I}}$ of $\mathrm{Sym}^{(\ell)}$, we get as coefficients what can be called the level ℓ monomial quasi-symmetric functions M $M_I(\mathbb{X})$

$$K(\mathbb{X}, \boldsymbol{A}) = \sum_{\mathbf{I}} M_{\mathbf{I}}(\mathbb{X}) S^{\mathbf{I}}(\boldsymbol{A}) \qquad (62)$$

defined by

$$M_{\mathbf{I}}(\mathbb{X}) = \sum_{j_1<\cdots<j_m} \mathbf{x}_{j_1}^{\mathbf{i}_1} \cdots \mathbf{x}_{j_m}^{\mathbf{i}_m} \,. \qquad (63)$$

with $I = (i_1, \cdots, i_m)$.

These functions form a basis of a subalgebra $\mathrm{QSym}^{(\ell)}$ of $K[\mathbb{X}]$, which we shall call the algebra of quasi-symmetric functions of level ℓ.

Poirier's Quasi-symmetric Functions

The functions $M_I(\mathbb{X})$ can be recognized as a basis of one of the algebras introduced by Poirier: the M_I coincide with the $M(C,v)$ defined in [34, p. 324, formula (1)], up to indexation.

Following Poirier, we introduce the level ℓ quasi-ribbon functions by summing over an order on ℓ-compositions: an ℓ-composition C is finer than C', and we write $C \geq C'$, if C' can be obtained by repeatedly summing up two consecutive columns of C such that no nonzero element of the left one is strictly below a nonzero element of the right one.

This order can be described in a much easier and natural way if one recodes an ℓ-composition I as a pair of words, the first one $d(I)$ being the set of sums of the elements of the first k columns of I, the second one $c(I)$ being obtained by concatenating the words $_i I_{i,j}$ while reading I by columns, from top to bottom and from left to right. For example, the 3-composition of Eq. (44) satisfies

$$d(\mathbf{I}) = \{5, 10, 14, 19\} \quad \text{and} \quad c(\mathbf{I}) = 13333\,22233\,1123\,12333. \tag{64}$$

Moreover, this recoding is a bijection if the two words $d(I)$ and $c(I)$ are such that the descent set of $c(I)$ is a subset of $d(I)$. The order previously defined on ℓ-compositions is in this context the inclusion order on sets d: $(d',c) \leq (d,c)$ iff $d' \subseteq d$.

It allows us to define the level ℓ quasi-ribbon functions F_I by

$$F_{\mathbf{I}} = \sum_{\mathbf{I}' \geq \mathbf{I}} M_{\mathbf{I}'} \tag{65}$$

Notice that this last description of the order $\leq$ is reminiscent of the reverse refinement order $\leq'$ on descent sets used in the context of quasi-symmetric functions and non-commutative symmetric functions: more precisely, since elements with different $c(I)$ are not comparable, the order $\geq$ restricted to ℓ -compositions of weight n is isomorphic to a direct sum of N of the order $\geq'$, where N is the number of possible words $c(I)$. The computation of its Möbius function is therefore straightforward.

Moreover, one can obtain the F_I as the commutative image of certain $F_{\sigma,u}$: any pair (σ, u) such that σ has descent set $d(I)$ and $u=c(I)$ will do.

Coproducts and Alphabets

Recall that to define the $G_{\sigma,u}(\mathbf{A})$ of an ℓ -colored alphabet $\mathbf{A} = A \times C$, we only need a total order on A, so that, if B is another copy of A commuting with A, we can define $\mathbf{A}+\mathbf{B}$ as $(A+B)\times C$, where $A+B$ is the ordinal sum, and thus make sense of $G_{\sigma,u}(\mathbf{A}+\mathbf{B})$. If, as usual, we identify $F(\mathbf{A})G(\mathbf{B})$ with $F\otimes G$, the coproduct defined by (17) is

$$\Delta F = \mathbf{F}(\boldsymbol{A} + \boldsymbol{B}) \tag{66}$$

Let us observe that on this picture, it is clear that the restriction of Δ to $\mathrm{Sym}^{(\ell)}$ is dual to the product of $\mathrm{QSym}^{(\ell)}$. By definition of the Cauchy kernel (61), we have

$$K(\mathbb{X}, \boldsymbol{A} + \boldsymbol{B}) = K(\mathbb{X}, \boldsymbol{A})K(\mathbb{X}, \boldsymbol{B}) \tag{67}$$

and by (62), this implies that Δ is dual to the multiplication of $\mathrm{QSym}^{(\ell)}$.

The same description can be applied to the quasi-symmetric side. Let

$\mathbb{X} = X \times C$ and $\mathbb{Y} = Y \times C$, where Y is a copy of X. Again, define X+Y as the ordinal sum of X and Y, and $\mathbb{X} + \mathbb{Y} = (X+Y) \times C$.

Lemma 5.1

The map $\nabla: F \mapsto F(\mathbb{X} + \mathbb{Y})$ defines a coproduct on $QSym^{(\ell)}$, which is dual to the product of $Sym^{(\ell)}$.

Proof

This follows from the identity

$$K(\mathbb{X} + \mathbb{Y}, \boldsymbol{A}) = K(\mathbb{X}, \boldsymbol{A}) K(\mathbb{Y}, \boldsymbol{A}). \tag{68}$$

The Internal Coproduct

From now on, we assume that the color set C is an additive semigroup, such that every element $\gamma \in C$ has a finite number of decompositions $\gamma = \alpha + \beta$.

We define the C-product $\mathbb{T} = \mathbb{X} \times_C \mathbb{Y}$ of two C-colored alphabets as the family of polynomials

$$\mathbb{T}^{(\gamma)} = \left\{ t_{rs}^{(\gamma)} = \sum_{\alpha+\beta=\gamma} x_r^{(\alpha)} y_s^{(\beta)} \right\} \tag{69}$$

with the pairs (r,s) ordered lexicographically.

Proposition 5.2

The map $\delta: F \mapsto F(\mathbb{X} \times_C \mathbb{Y})$ defines a coassociative coproduct on $QSym^{(\ell)}$. We define the internal product $*$ of $Sym^{(\ell)}$ as the product dual to δ.

Proof

The coassociativity condition

$$(\delta \otimes \mathrm{Id}) \circ \delta = (\mathrm{Id} \otimes \delta) \circ \delta \tag{70}$$

translates as the associativity of the C-product

$$(\mathbb{X} \times_C \mathbb{Y}) \times_C \mathbb{Z} = \mathbb{X} \times_C (\mathbb{Y} \times_C \mathbb{Z}) \tag{71}$$

which is clear since both sides are by definition

$$\left\{ t^{\mu}_{pqr} = \sum_{\alpha+\beta+\gamma=\mu} x_p^{(\alpha)} y_q^{(\beta)} z_r^{(\gamma)} \right\} \tag{72}$$

with the lexicographic order on triples (p,q,r).

The Splitting Formula

The definition of $*$ by duality with the C-product implies that it satisfies the splitting formula. We just need to check a trivial property:

Lemma 5.3

The C-product is distributive over the colored ordinal sum:

$$(\mathbb{X}' + \mathbb{X}'') \times_C \mathbb{Y} = \mathbb{X}' \times_C \mathbb{Y} + \mathbb{X}'' \times_C \mathbb{Y}. \tag{73}$$

Proposition 5.4

Let $\mu_r : \left(Sym^{(\ell)}\right)^{\otimes r} \to Sym^{(\ell)}$ be the product map. Let $\Delta^{(r)} : \left(Sym^{(\ell)}\right) \to \left(Sym^{(\ell)}\right)$ be the r-fold coproduct, and $_{*r}$ be the extension of the internal product to $\left(Sym^{(\ell)}\right)^{\otimes r}$. Then, for $F_1, \ldots, F_r$, and $G \in Sym^{(\ell)}$,

$$(F_1 \cdots F_r) * G = \mu_r[(F_1 \otimes \cdots \otimes F_r) *_r \Delta^{(r)} G] \tag{74}$$

Proof

It is enough to consider the case r=2. Let $F \in QSym^{(\ell)}$ and $U, W, W \in Sym^{(\ell)}$. We have, writing $\mathbb{XY} = \mathbb{X} \times_C \mathbb{Y}$ for short, and assuming duality between $\mathbb{X}$ and A, $\mathbb{Y}$ and $\boldsymbol{B}$ and so on,

$$\begin{aligned}
\langle F, (UV) * W\rangle &= \langle F(\mathbb{X}'\mathbb{X}''), (UV)(\boldsymbol{A}')W(\boldsymbol{A}'')\rangle \\
&= \langle F((\mathbb{X}' + \mathbb{Y}')\mathbb{X}''), U(\boldsymbol{A}')V(\boldsymbol{B}')W(\boldsymbol{A}'')\rangle \\
&= \langle F(\mathbb{X}'\mathbb{X}'' + \mathbb{Y}'\mathbb{Y}''), U(\boldsymbol{A}')V(\boldsymbol{B}')W(\boldsymbol{A}'' + \boldsymbol{B}'')\rangle \\
&= \langle F(\mathbb{X}' + \mathbb{Y}'), U(\boldsymbol{A}')V(\boldsymbol{B}') *_2 W(\boldsymbol{A}' + \boldsymbol{B}')\rangle \\
&= \langle \nabla F, (U \otimes V) *_2 \Delta W\rangle = \langle F, \mu[(U \otimes V) *_2 \Delta W]\rangle .
\end{aligned}$$

The third expression follows from the second one by expanding $F((\mathbb{X}' + \mathbb{Y}')\mathbb{X}'')$ as a sum of products of functions of X''

$$F((\mathbb{X}' + \mathbb{Y}')\mathbb{X}'') = \sum_{(F)} F_{(1)}(\mathbb{X}'\mathbb{X}'')F_{(2)}(\mathbb{Y}'\mathbb{X}'') \tag{75}$$

and dualizing the product w.r.t. $\mathbb{X}''$, which results in replacing $\mathbb{X}''$ by $\mathbb{Y}''$ in the second factor of each term of the sum, which yields

$$\sum_{(F)} F_{(1)}(\mathbb{X}'\mathbb{X}'')F_{(2)}(\mathbb{Y}'\mathbb{Y}'') = F(\mathbb{X}'\mathbb{X}'' + \mathbb{Y}'\mathbb{Y}'') \tag{76}$$

while on the right part of the bracket, the dual alphabet A'' is replaced by $A''+B''$.

The fourth line follows from the third one by dualizing $(\mathbb{X}'\mathbb{X}'',A'')\rightarrow(\mathbb{X}',A')$ and $(\mathbb{Y}'\mathbb{Y}'',B'')\rightarrow(\mathbb{Y}',B')$ which results in an internal product in the right hand-side of the bracket.

Evaluation of Internal Products

Let us start with the simplest case, $S_i * S_j$. The coefficient of S^K in this product is the coefficient of $M_i \otimes M_j$ in δM_K, which is also the coefficient of $x^i y^j$ in $M_K(x \times cy)$. This is zero if **K** has more than one column, so that

$$S_{\mathbf{i}} * S_{\mathbf{j}} = \sum_{\mathbf{n}} d_{\mathbf{ij}}^{\mathbf{n}} S_{\mathbf{n}} \tag{77}$$

contains only one-part vector compositions.

Lemma 5.5

If the color semigroup is $C=N$, then the coefficient d_{ij}^n of S_n in $S_i * S_j$ is equal to the coefficient of the monomial symmetric function m_μ in the product $m_\alpha m_\beta$, where μ is the partition $(0^{n_0}1^{n_1}\dots)$, $\alpha=(0^{i_0}1^{i_1}\dots)$, $\beta=(0^{j_0}1^{j_1}\dots)$, the monomial functions being taken over an alphabet of n letters, where $n=|i|$.

Proof

From (77), we only need to compute the coproducts δM_n. For this, it is sufficient to use alphabets of the form $x=\{x\}\times C, y=\{y\}\times C$. Then,

$$M_{\mathbf{n}}(\mathbf{x}\times_C \mathbf{y}) = \prod_{k\geq 0}\left(\sum_{i+j=k} x^{(i)}y^{(j)}\right)^{n_k} \tag{78}$$

and we see that the coefficient of $M_i(x)M_j(Y) = x^i y^j$ is the same as the coefficient of $h_\alpha \otimes h_\beta$ in

$$\prod_{k\geq 0}\left(\sum_{i+j=k} h_i \otimes h_j\right)^{n_k} = \Delta\left(h_0^{n_0} h_1^{n_1} \cdots\right) = \Delta h_\mu . \quad \square \tag{79}$$

For example,

$$S^{\begin{pmatrix}2\\2\end{pmatrix}} * S^{\begin{pmatrix}3\\1\end{pmatrix}} = 3S^{\begin{pmatrix}1\\3\end{pmatrix}} + S^{\begin{pmatrix}2\\1\\1\end{pmatrix}} \tag{80}$$

This result is compatible with the fact that

$$m_{11}m_1 = 3m_{111} + m_{21} \tag{81}$$

As another example,

$$S^{\begin{pmatrix}0\\2\\1\end{pmatrix}} * S^{\begin{pmatrix}1\\1\\1\end{pmatrix}} = S^{\begin{pmatrix}0\\1\\1\\0\\1\end{pmatrix}} + 2S^{\begin{pmatrix}0\\1\\0\\2\end{pmatrix}} + 2S^{\begin{pmatrix}0\\0\\2\\1\end{pmatrix}} \tag{82}$$

One can then check that the previous result amounts to selecting the partitions of length at most 3 in

$$m_{211}m_{21} = m_{421} + 2m_{331} + 2m_{322} + \cdots \tag{83}$$

Together with the splitting formula (74), this result determines all the products $S^I * S^J$, since one also has

$$S_{\mathbf{i}} * S^{\mathbf{J}} = S^{\mathbf{J}} * S_{\mathbf{i}} \tag{84}$$

thanks to the isomorphism of ordered colored alphabets

$$\mathbf{X} \times_C \mathbf{Y} \simeq \mathbf{Y} \times_C \mathbf{X} \tag{85}$$

where $x = \{x\} \times \otimes C$.

When the color group is $Z/\ell Z$, the result is obtained by reduction modulo ℓ of the colors, e.g. , with $\ell = 3$, we get from example (82)

$$S^{\begin{pmatrix}0\\2\\1\end{pmatrix}} * S^{\begin{pmatrix}1\\1\\1\end{pmatrix}} = 2S^{\begin{pmatrix}0\\2\\1\end{pmatrix}} + 2S^{\begin{pmatrix}2\\1\\0\end{pmatrix}} + 2S^{\begin{pmatrix}1\\0\\2\end{pmatrix}} \tag{86}$$

Note that the coefficient of an S_I can change when computing modulo ℓ: for all pairs of parts k and k′ added together, a normalization factor $\binom{k+k'}{k}$ appears because of definition (45).

Example 5.6

With $\ell = 2$ and $C = Z/2Z$,

$$\begin{aligned}
S^{\begin{pmatrix}1&0\\1&1\end{pmatrix}} * S^{\begin{pmatrix}0&2\\1&0\end{pmatrix}} &= \mu_2\left[\left(S^{\begin{pmatrix}1\\1\end{pmatrix}} \otimes S^{\begin{pmatrix}0\\1\end{pmatrix}}\right) *_2 \Delta S^{\begin{pmatrix}0&2\\1&0\end{pmatrix}}\right] \\
&= \left(S^{\begin{pmatrix}1\\1\end{pmatrix}} * S^{\begin{pmatrix}2\\0\end{pmatrix}}\right)\left(S^{\begin{pmatrix}0\\1\end{pmatrix}} * S^{\begin{pmatrix}0\\1\end{pmatrix}}\right) + \left(S^{\begin{pmatrix}1\\1\end{pmatrix}} * S^{\begin{pmatrix}0&1\\1&0\end{pmatrix}}\right)\left(S^{\begin{pmatrix}0\\1\end{pmatrix}} * S^{\begin{pmatrix}1\\0\end{pmatrix}}\right) \\
&= S^{\begin{pmatrix}1&1\\1&0\end{pmatrix}} + S^{\begin{pmatrix}1&1&0\\0&0&1\end{pmatrix}} + S^{\begin{pmatrix}0&0&0\\1&1&1\end{pmatrix}}.
\end{aligned}$$

Generalized Descent Algebras

In a preliminary draft of this work [30], we introduced the internal product in a different way, similar to Solomon's original definition of the descent algebra. We shall now show that both definitions coincide. Assuming that C is a commutative monoid, we regard colored permutations as elements of the wreath product $H = C \wr S_n$, so that we can consider the restriction to $Sym^{(c)}$ of the internal product defined by (43), provisionally denoted by $*'$: for $h', h'' \in H$,

$$\mathbf{G}_{h'} *' \mathbf{G}_{h''} = \mathbf{G}_{h''h'} \tag{87}$$

(opposite law, as in the classical case of Sym).

Theorem 5.7

Let C be a commutative (additive) monoid.

1. $Sym_n^{(C)}$ is a subalgebra of $FQSym_n^{(C)}$, for the operation $*'$.
2. The restriction of $*'$ to $Sym_n^{(C)}$ satisfies the splitting formula (74).
3. $S_{\mathbf{i}} *' S^{\mathbf{J}} = S^{\mathbf{J}} *' S_{\mathbf{i}}$
4. $S_i *' S_j = S^J * S_i$, so that $*'$ coincides with $*$.

Proof

The proofs of (1) and (2) can be found in [1]. (3) and (4) follow from the definition in (45) and the internal product on $FQSym^{(C)}$ given by (39). Indeed, the S_i are clearly central, and when $C = N$, $S_i *' S_j$ is the sum of all colorings of the identity $w = u + v$ such that $|u|_c = i_c$ and $|v|_c = j_c$ for all c.

This provides an analogue of Solomon's descent algebra for the wreath product $C \wr S_n$. Note that the definition remains valid for C=Z, so that we get a descent algebra for the (extended) affine Weyl groups of type A, $\hat{\mathfrak{S}}_n = Z \wr \mathfrak{S}_n$.

Ordinary Multi-symmetric Functions

Recall that if one sets $X_i=\{x_{i,j} \mid j=1,\ldots,n\}$, the multi-symmetric polynomials are the invariants of the symmetric group $\mathfrak{S}_n$ in $C[X_0,\ldots,X_{\ell-1}]$ for the diagonal action (by the automorphisms $\sigma(x_{i,j})=x_{i,\sigma(j)}$). The Hopf algebra $Sym^{(\ell)}$ of multi-symmetric functions of level ℓ is the inverse limit of this sequence in the category of graded algebras for $n\to\infty$ [27] and [8].

A consequence of the above results is that $Sym^{(\ell)}$ admits an internal product. If we denote by $\underline{F}\in Sym^{(\ell)}$ the commutative image of $F\in Sym^{(\ell)}$, obtained by sending the noncommuting variables $a_j^{(i)}$ to the commuting $x_{i,j}=x_j^{(i)}$, the map $F\mapsto\underline{F}$ is an epimorphism of Hopf algebras (the generators S_n are mapped onto the generators h_n of Sym^{ℓ}), and we have

$$\underline{F*G}=\underline{F'*G'} \quad \text{as soon as } \underline{F}=\underline{F'} \text{ and } \underline{G}=\underline{G'} \tag{88}$$

Indeed, the splitting formula (74) implies that $F*G=F'*G$ for $\underline{F}=\underline{F'}$, and the definition of $*$ as dual to δ implies that $\underline{F*G}=\underline{G*F}$.

For example,

$$\begin{aligned}
S^{\begin{pmatrix}0&3\\1&1\end{pmatrix}} * S^{\begin{pmatrix}1&2\\1&1\end{pmatrix}} &= S^{\begin{pmatrix}0&0&2\\1&0&1\\0&1&0\end{pmatrix}} + S^{\begin{pmatrix}0&0&2\\0&1&1\\1&0&0\end{pmatrix}} + S^{\begin{pmatrix}0&1&2\\0&0&0\\1&0&1\end{pmatrix}} + S^{\begin{pmatrix}0&0&2\\1&1&0\\0&0&1\end{pmatrix}}\\
&+ 2S^{\begin{pmatrix}0&1&1\\0&0&2\\1&0&0\end{pmatrix}} + 2S^{\begin{pmatrix}0&0&1\\1&1&2\\0&0&0\end{pmatrix}} + S^{\begin{pmatrix}1&0&1\\0&1&1\\1&0&0\end{pmatrix}} + S^{\begin{pmatrix}1&0&1\\1&1&0\\0&0&1\end{pmatrix}} + S^{\begin{pmatrix}1&0&1\\1&0&1\\0&1&0\end{pmatrix}}\\
&+ S^{\begin{pmatrix}1&0&2\\0&0&0\\1&1&0\end{pmatrix}} + 2S^{\begin{pmatrix}1&0&0\\1&1&2\\0&0&0\end{pmatrix}} + 2S^{\begin{pmatrix}0&0&1\\2&1&1\\0&0&0\end{pmatrix}} + 2S^{\begin{pmatrix}0&0&2\\2&0&0\\0&1&0\end{pmatrix}}
\end{aligned} \tag{89}$$

$$S^{\begin{pmatrix}0&3\\1&1\end{pmatrix}} * S^{\begin{pmatrix}2&1\\1&1\end{pmatrix}} = S^{\begin{pmatrix}2&0&0\\1&0&1\\0&1&0\end{pmatrix}} + S^{\begin{pmatrix}2&0&0\\1&1&0\\0&0&1\end{pmatrix}} + S^{\begin{pmatrix}2&0&1\\0&0&0\\1&1&0\end{pmatrix}} + S^{\begin{pmatrix}2&0&0\\0&1&1\\1&0&0\end{pmatrix}}$$
$$+ 2S^{\begin{pmatrix}1&0&1\\2&0&0\\0&1&0\end{pmatrix}} + 2S^{\begin{pmatrix}1&0&0\\2&1&1\\0&0&0\end{pmatrix}} + S^{\begin{pmatrix}0&1&1\\1&0&1\\0&1&0\end{pmatrix}} + S^{\begin{pmatrix}0&1&1\\1&1&0\\0&0&1\end{pmatrix}} + S^{\begin{pmatrix}0&1&1\\0&1&1\\1&0&0\end{pmatrix}}$$
$$+ S^{\begin{pmatrix}0&2&1\\0&0&0\\1&0&1\end{pmatrix}} + 2S^{\begin{pmatrix}0&0&1\\1&2&1\\0&0&0\end{pmatrix}} + 2S^{\begin{pmatrix}0&1&0\\1&1&2\\0&0&0\end{pmatrix}} + 2S^{\begin{pmatrix}0&2&0\\0&0&2\\1&0&0\end{pmatrix}} \tag{90}$$

$$S^{\begin{pmatrix}3&0\\1&1\end{pmatrix}} * S^{\begin{pmatrix}1&2\\1&1\end{pmatrix}} = S^{\begin{pmatrix}0&0&2\\1&0&1\\0&1&0\end{pmatrix}} + S^{\begin{pmatrix}0&0&2\\0&1&1\\1&0&0\end{pmatrix}} + S^{\begin{pmatrix}1&0&2\\0&0&0\\0&1&1\end{pmatrix}} + S^{\begin{pmatrix}0&0&2\\1&1&0\\0&0&1\end{pmatrix}}$$
$$+ 2S^{\begin{pmatrix}1&0&1\\0&0&2\\0&1&0\end{pmatrix}} + 2S^{\begin{pmatrix}0&0&1\\1&1&2\\0&0&0\end{pmatrix}} + S^{\begin{pmatrix}1&1&0\\0&1&1\\1&0&0\end{pmatrix}} + S^{\begin{pmatrix}1&1&0\\1&0&1\\0&1&0\end{pmatrix}} + S^{\begin{pmatrix}1&1&0\\1&1&0\\0&0&1\end{pmatrix}}$$
$$+ S^{\begin{pmatrix}1&2&0\\0&0&0\\1&0&1\end{pmatrix}} + 2S^{\begin{pmatrix}1&0&0\\1&2&1\\0&0&0\end{pmatrix}} + 2S^{\begin{pmatrix}0&1&0\\2&1&1\\0&0&0\end{pmatrix}} + 2S^{\begin{pmatrix}0&2&0\\2&0&0\\0&0&1\end{pmatrix}} \tag{91}$$

$$S^{\begin{pmatrix}3&0\\1&1\end{pmatrix}} * S^{\begin{pmatrix}2&1\\1&1\end{pmatrix}} = S^{\begin{pmatrix}2&0&0\\1&0&1\\0&1&0\end{pmatrix}} + S^{\begin{pmatrix}2&0&0\\1&1&0\\0&0&1\end{pmatrix}} + S^{\begin{pmatrix}2&1&0\\0&0&0\\1&0&1\end{pmatrix}} + S^{\begin{pmatrix}2&0&0\\0&1&1\\1&0&0\end{pmatrix}}$$
$$+ 2S^{\begin{pmatrix}1&1&0\\2&0&0\\0&0&1\end{pmatrix}} + 2S^{\begin{pmatrix}1&0&0\\2&1&1\\0&0&0\end{pmatrix}} + S^{\begin{pmatrix}1&0&1\\0&1&1\\1&0&0\end{pmatrix}} + S^{\begin{pmatrix}1&0&1\\1&1&0\\0&0&1\end{pmatrix}} + S^{\begin{pmatrix}1&0&1\\1&0&1\\0&1&0\end{pmatrix}}$$
$$+ S^{\begin{pmatrix}2&0&1\\0&0&0\\0&1&1\end{pmatrix}} + 2S^{\begin{pmatrix}0&0&1\\2&1&1\\0&0&0\end{pmatrix}} + 2S^{\begin{pmatrix}1&0&0\\1&1&2\\0&0&0\end{pmatrix}} + 2S^{\begin{pmatrix}2&0&0\\0&0&2\\0&1&0\end{pmatrix}}. \tag{92}$$

If one denotes by h the commutative image of S, one easily checks that

$$h^{\begin{pmatrix}3\\1\end{pmatrix}} h^{\begin{pmatrix}0\\1\end{pmatrix}} * h^{\begin{pmatrix}2\\1\end{pmatrix}} h^{\begin{pmatrix}1\\1\end{pmatrix}} = 2h^{\begin{pmatrix}2\\1\\0\end{pmatrix}} h^{\begin{pmatrix}0\\0\\1\end{pmatrix}} h^{\begin{pmatrix}0\\1\\0\end{pmatrix}} + h^{\begin{pmatrix}2\\0\\1\end{pmatrix}} h^{\begin{pmatrix}1\\0\\0\end{pmatrix}} h^{\begin{pmatrix}0\\0\\1\end{pmatrix}} + h^{\begin{pmatrix}2\\0\\1\end{pmatrix}} h^{\begin{pmatrix}0\\1\\0\end{pmatrix}} h^{\begin{pmatrix}0\\1\\0\end{pmatrix}}$$
$$+ 2h^{\begin{pmatrix}1\\2\\0\end{pmatrix}} h^{\begin{pmatrix}1\\0\\0\end{pmatrix}} h^{\begin{pmatrix}0\\0\\1\end{pmatrix}} + 2h^{\begin{pmatrix}1\\2\\0\end{pmatrix}} h^{\begin{pmatrix}0\\1\\0\end{pmatrix}} h^{\begin{pmatrix}0\\1\\0\end{pmatrix}}$$
$$+ 2h^{\begin{pmatrix}1\\1\\0\end{pmatrix}} h^{\begin{pmatrix}1\\0\\1\end{pmatrix}} h^{\begin{pmatrix}0\\1\\0\end{pmatrix}} + h^{\begin{pmatrix}1\\1\\0\end{pmatrix}} h^{\begin{pmatrix}1\\1\\0\end{pmatrix}} h^{\begin{pmatrix}0\\0\\1\end{pmatrix}}$$
$$+ h^{\begin{pmatrix}2\\0\\0\end{pmatrix}} h^{\begin{pmatrix}1\\0\\1\end{pmatrix}} h^{\begin{pmatrix}0\\0\\1\end{pmatrix}} + 4h^{\begin{pmatrix}1\\1\\0\end{pmatrix}} h^{\begin{pmatrix}0\\2\\0\end{pmatrix}} h^{\begin{pmatrix}0\\1\\0\end{pmatrix}} + 2h^{\begin{pmatrix}2\\0\\0\end{pmatrix}} h^{\begin{pmatrix}0\\2\\0\end{pmatrix}} h^{\begin{pmatrix}0\\0\\1\end{pmatrix}}. \tag{93}$$

THE MANTACI–REUTENAUER ALGEBRA

Monochromatic Complete Functions

Let e_i be the canonical basis of N^{ℓ}. For $n \geq 1$, let

$$S_n^{(i)} = S_{n \cdot \mathbf{e}_i} \in \mathbf{Sym}^{(\ell)} \tag{94}$$

be the monochromatic complete symmetric functions. For a colored composition $(I,u)=((i_1,\ldots,i_m),(u_1,\ldots,u_m))$, we set

$$S^{(I,u)} = S_{i_1}^{(u_1)} \cdots S_{i_m}^{(u_m)} \tag{95}$$

Proposition 6.1

The $S_n^{(i)}$ generate a Hopf subalgebra $MR^{(\ell)}$ of $Sym^{(\ell)}$, which is isomorphic to the Mantaci–Reutenauer descent algebra of the wreath products $\mathfrak{S}_i^{\ell} = (Z/\ell Z \wr \mathfrak{S}_n)$ if $C=Z/\ell Z$.

Proof

$MR^{(\ell)}$ is obviously stable by the product and coproduct coming from $Sym^{(\ell)}$, hence is a Hopf subalgebra of $Sym^{(\ell)}$.

As a Hopf algebra, it is clearly isomorphic to the Mantaci–Reutenauer algebra, having the same number of independent generators in each degree, and behaving in the same way (divided powers) under the coproduct. The isomorphism for the internal product comes from the splitting formula, which gives explicitly$S^{(I,u)} * S^{(J,v)}$.

Since $MR^{(\ell)}$ has ℓ generators in each degree, its dimensions are given by

$$\frac{1}{1-\sum_{k\geq 1}\ell t^k} = \frac{1-t}{1-(\ell+1)t} = 1+\ell\sum_{n\geq 1}(\ell+1)^{n-1}t^n \tag{96}$$

The bases of $MR^{(\ell)}$ are labeled by colored compositions.

Primitive Elements of $MR^{(\ell)}$

$MR^{(\ell)}$ being a subalgebra of a graded connected cocommutative Hopf algebra, is itself a graded connected cocommutative Hopf algebra, so that, thanks to the Cartier–Milnor–Moore theorem, it is the universal enveloping algebra of $L^{(\ell)}$:

$$MR^{(\ell)} = U(L^{(\ell)}) \tag{97}$$

where $L^{(\ell)}$ is the Lie algebra of its primitive elements. Since $MR^{(\ell)}$ is a free associative algebra with ℓ generators in each degree, we have, by the usual argument as in the proof of Theorem 4.2:

Theorem 6.2

As a graded Lie algebra, the primitive Lie algebra $L^{(\ell)}$ of $MR^{(\ell)}$ is free over a set of ℓ generators in each degree.

For example, with $\ell=2$, the generating series of the dimensions of $L^{(\ell)}$ is

$$2t+3t^2+8t^3+18t^4+48t^5+116t^6+312t^7+\cdots \tag{98}$$

With $\ell=3$, one finds

$$3t + 6t^2 + 20t^3 + 60t^4 + 204t^5 + 670t^6 + 2340t^7 + \cdots \tag{99}$$

The dimension of $L_n^{(\ell)}$ is given by the Witt polynomials

$$q_n(\ell) := \begin{cases} \ell & \text{if } n = 1, \\ \dfrac{1}{n}\sum_{d|n} \mu(d)(\ell+1)^{n/d} & \text{if } n > 1. \end{cases} \tag{100}$$

Note that, for $n \geq 2$, the dimension of $L_n^{(\ell)}$ coincide with those of a free Lie algebra on $\ell + 1$ generators of degree 1. Indeed, writing

$$\prod_{n\geq 1}(1 - t^n)^{-q_n(\ell)} = \frac{1-t}{1-(\ell+1)t} \tag{101}$$

and taking logarithms, we obtain (100) by Möbius inversion.

Duality

The duality is easily worked out by means of the appropriate Cauchy kernel. The generating function of the complete functions is

$$\sigma_{\mathbf{x}}^{\mathrm{MR}}(\boldsymbol{A}) := 1 + \sum_{j=0}^{\ell-1}\sum_{n\geq 1} S_n^{(j)} \cdot (x^{(j)})^n \tag{102}$$

and the Cauchy kernel is as usual

$$K^{\mathrm{MR}}(\mathbb{X}, \boldsymbol{A}) = \overrightarrow{\prod_{i\geq 1}} \sigma_{\mathbf{x}_i}^{\mathrm{MR}}(\boldsymbol{A}) = \sum_{(I,u)} M_{(I,u)}(\mathbb{X}) S^{(I,u)}(\boldsymbol{A}) \tag{103}$$

where (I, u) runs over colored compositions $(I, u) = ((i_1, \ldots, i_m), (u_1, \ldots, u_m))$. The $M_{I,u}$ are called themonochromatic monomial quasi-symmetric functions and satisfy

$$M_{(I,u)}(\mathbb{X}) = \sum_{j_1<\cdots<j_m} (x_{j_1}^{(u_1)})^{i_1} \cdots (x_{j_m}^{(u_m)})^{i_m} \quad (104)$$

Proposition 6.3

The $M_{(I,u)}$ span a subalgebra of $C[X]$ which can be identified with the graded dual of $MR^{(\ell)}$ through the pairing

$$\langle M_{(I,u)}, S^{(J,v)} \rangle = \delta_{I,J}\delta_{u,v} \quad (105)$$

where δ is the Kronecker symbol.

Note that this algebra can also be obtained by imposing the relations

$$x_i^{(p)} x_i^{(q)} = 0, \quad \text{for } p \neq q \quad (106)$$

on the variables of $QSym^{(\ell)}$.

Baumann and Hohlweg [1] have another construction of the dual of $MR^{(\ell)}$ (implicitly defined in [34, Lemma 11]).

LEVEL ℓ PARKING QUASI-SYMMETRIC FUNCTIONS

Usual Parking Functions

The Combinatorial Objects

Recall that a parking function on $[n]=\{1,2,\ldots,n\}$ is a word $a = a_1a_2\cdots a_n$ of length n on $[n]$ whose nondecreasing rearrangement $a^{\uparrow} = a'_1a'_2\cdots a'_n$ satisfies $a'_i \leq i$ for all i. Let PF_n be the set of such words. It is well-known that $|PF_n| = (n+1)^{n-1}$.

One says that a has a breakpoint at b if $|\{a_i \leq b\}| = b$. The set of all breakpoints of a is denoted by $BP(a)$. Then, $a \in PF_n$ is said to be prime if $BP(a) = \{n\}$ (see [41]).

Let $PPF_n \subset PF_n$ be the set of prime parking functions on $[n]$. It can easily be shown that $|PPF_n| = (n-1)^{n-1}$ (see [42]).

Finally, one says that a has a match at b if $|\{a_i < b\}| = b - 1$ and $|\{a_i \leq b\}| \geq b$. The set of all matches of a is denoted by $Ma(a)$.

Algebraic Structure on Parking Functions

The algebra PQSym of parking functions [29] and [31] is very similar to the algebra FQSym of permutations.

Since parking functions are closed under the shifted shuffle, one defines a product on the vector space with basis (F_a) by

$$\mathbf{F}_{\mathbf{a}'}\mathbf{F}_{\mathbf{a}''} = \sum_{\mathbf{a} \in \mathbf{a}' \sqcup\!\sqcup\, \mathbf{a}''[k]} \mathbf{F}_{\mathbf{a}}. \tag{107}$$

The coproduct on this basis is given by the parkization algorithm [29]: for $w = w_1 w_2 \cdots w_n$ on $\{1, 2, \ldots\}$, let us define

$$d(w) := \min\{i | \#\{w_j \leq i\} < i\} \tag{108}$$

If $d(w) = n+1$, then w is a parking function and the algorithm terminates, returning w. Otherwise, let w' be the word obtained by decrementing all the elements of w greater than d(w). Then Park(w) := Park(w′). Since w′ is smaller than w in the lexicographic order, the algorithm terminates and always returns a parking function.

For example, let w=(3,5,1,1,11,8,8,2). Then d(w)=6 and the word w′=(3,5,1,1,10,7,7,2). Then d(w′)=6 and w″=(3,5,1,1,9,6,6,2). Finally, d(w″)=8 and w‴=(3,5,1,1,8,6,6,2), that is a parking function. Thus, Park(w) = (3,5,1,1,8,6,6,2).

We then have

$$\Delta \mathbf{F}_{\mathbf{a}} = \sum_{\substack{u,v \\ u\cdot v=\mathbf{a}}} \mathbf{F}_{\mathrm{Park}(u)} \otimes \mathbf{F}_{\mathrm{Park}(v)} \tag{109}$$

Duality

Let $G_a = F_a^* \in \mathrm{PQSym}^*$ denote the dual basis of (F_a). Note that at this point, the G_a have nothing to do with those of Section 3.1. If $\langle , \rangle$ denotes the duality bracket, the product on PQSym^* is given by

$$\mathbf{G}_{\mathbf{a}'}\mathbf{G}_{\mathbf{a}''} = \sum_{\mathbf{a}} \langle \mathbf{G}_{\mathbf{a}'} \otimes \mathbf{G}_{\mathbf{a}''}, \Delta \mathbf{F}_{\mathbf{a}} \rangle \mathbf{G}_{\mathbf{a}} = \sum_{\mathbf{a} \in \mathbf{a}' * \mathbf{a}''} \mathbf{G}_{\mathbf{a}}, \tag{110}$$

where the convolution $a' * a''$ of two parking functions is defined as

$$\mathbf{a}' * \mathbf{a}'' = \sum_{\substack{u,v;\, \mathbf{a}=u\cdot v, \\ \mathrm{Park}(u)=\mathbf{a}', \mathrm{Park}(v)=\mathbf{a}''}} \mathbf{a}. \tag{111}$$

By duality, one easily gets the formula for the coproduct of G_a as

$$\Delta \mathbf{G_a} := \sum_{u,v;\mathbf{a}\in u \Cup v} \mathbf{G}_u \otimes \mathbf{G}_v . \tag{112}$$

Given a word w on $\{1,2,\ldots\}$, it is possible to define a notion of parkization Park(w), a parking function which coincides with Std(w) when w is a word without repeated letters [29]. Then, one has the following realization:

$$\mathbf{G_a}(A) := \sum_{w\in A^*, \mathrm{Park}(w)=\mathbf{a}} w . \tag{113}$$

Despite the fact that PQSym is self-dual, the realization of the F_a is not known in general.

Colored Parking Functions

Let ℓ be an integer representing the number of allowed colors. A colored parking function of level ℓ and size n is a pair composed of a parking function of length n and a word on $[\ell]$ of length n.

Since there is no restriction on the coloring, it is obvious that there are $\ell^n(n+1)^{n-1}$ colored parking functions of level ℓ and size n.

With two colors, one finds the sequence $a_i=2(2i+2)^{i-1}$, known as A097629 in [40]:

$$1 + 2t + 12t^2 + 128t^3 + 2000t^4 + 41472t^5 + 1\,075\,648t^6 + \cdots \tag{114}$$

Since the convolution of two parking functions contains only parking functions, one easily builds as in [29] an algebra $\mathrm{PQSym}^{(\ell)*}$ on colored parking functions:

$$\mathbf{G}_{(\mathbf{a}',u')}\mathbf{G}_{(\mathbf{a}'',u'')} = \sum_{\mathbf{a}\in\mathbf{a}'*\mathbf{a}''} \mathbf{G}_{(\mathbf{a},u'\cdot u'')} \tag{115}$$

We start from the dual side, since the realization of the F basis on words is not known in general.

We can now again define a coproduct using sums of alphabets: we only need a total order on A to define the colored parkization, so that taking two isomorphic copies A′ and A″ of A, and defining as above A′ + A″ as (A′+A″)×C, the coproduct is given by:

$$\Delta f(A) = f(A' + A'') \tag{116}$$

By construction, this is an algebra morphism from $\mathrm{PQSym}^{(\ell)*}$ to $\mathrm{PQSym}^{(\ell)*} \otimes \mathrm{PQSym}^{(\ell)*}$, so that

Theorem 7.1

$\mathrm{PQSym}^{(\ell)*}$ is a graded connected bialgebra, hence a Hopf algebra. More precisely, the coproduct can be computed in the following way:

$$\Delta\mathbf{G}_{(\mathbf{a},u)} = \sum_{\substack{(\mathbf{a}',\mathbf{a}'',u',u'') \\ (\mathbf{a},u)\in(\mathbf{a}',u')\Cup(\mathbf{a}'',u'')}} \mathbf{G}_{(\mathbf{a}',u')} \otimes \mathbf{G}_{(\mathbf{a}'',u'')} \tag{117}$$

Proof

Completely similar to the case of $\mathrm{FQSym}^{(\ell)}$.

For example,

$$\Delta\mathbf{G}_{(41142,22115)} = 1 \otimes \mathbf{G}_{(41142,22115)} + \mathbf{G}_{(112,215)} \otimes \mathbf{G}_{(11,21)} + \mathbf{G}_{(41142,22115)} \otimes 1 \tag{118}$$

Duality

We can now define $\mathrm{PQSym}^{(\ell)}$ as the graded dual of $\mathrm{PQSym}^{(\ell)*}$.

Let $F_{(a,u)} = G^{*}_{(a,u)} \in \mathrm{PQSym}^{(\ell)}$ be the dual basis of (G_a). If $\langle,\rangle$ denotes the duality bracket, the product on $\mathrm{PQSym}^{(\ell)}$ is given by

$$\mathbf{F}_{(\mathbf{a}',u')}\mathbf{F}_{(\mathbf{a}'',u'')} = \sum_{(\mathbf{a},u)\in(\mathbf{a}',u')\Cup(\mathbf{a}'',u'')} \mathbf{F}_{(\mathbf{a},u)}, \tag{119}$$

where the shifted shuffle of two colored parking functions is such that colors follow their letters.

Using the duality bracket once more, one easily gets the formula for the coproduct of $F_{(a,u)}$ as

$$\Delta\mathbf{F}_{(\mathbf{a},u)} = \sum_{\substack{(p',u'),(p'',u'')\\ p'p''=\mathbf{a};\,u'u''=u}} \mathbf{F}_{(\mathrm{Park}(p'),u')} \otimes \mathbf{F}_{(\mathrm{Park}(p''),u'')}. \tag{120}$$

Algebraic Structure of $\mathrm{PQSym}^{(\ell)}$

Recall that a word w over N^* is connected if it cannot be written as a shifted concatenation $w=u\bullet v$, andanti-connected if its mirror image $\overline{w}$ is connected.

We denote by CP the set of connected parking functions, and by $p_n := |CP_n|$ the number of such parking functions of size n. We recall that the generating series of p_n is Sequence A122708 of [40]:

$$p(t) := \sum_{n\geq 1} p_n t^n = 1 - \left(\sum_{n\geq 0}(n+1)^{n-1}t^n\right)^{-1}$$
$$= t + 2t^2 + 11t^3 + 92t^4 + 1014t^5 + 13\,795t^6 + 223\,061t^7 + \cdots.$$

Let the connected colored parking functions be the (a,u) with a connected and u arbitrary. Their generating series is given by $p(\ell t)$.

From [31], we immediately get

Proposition 7.2

$PQSym^{(\ell)}$ (resp. $PQSym^{(\ell)*}$) is free over the set $F_{\sigma,u}$ (resp. $G_{\sigma,u}$), where (σ,u) is connected.

For example, one gets the following generating series for the algebraic generators (connected parking functions with $\ell=2$):

$$2t + 8t^2 + 88t^3 + 1472t^4 + 32\,448t^5 + 882\,880t^6 + 28\,551\,808t^7 + \cdots. \tag{121}$$

Bidendriform Structure

Theorem 7.3

The algebra $PQSym^{(\ell)}$ has a structure of bidendriform bialgebra, hence is free as a Hopf algebra and as a dendriform algebra, cofree, self-dual, and its primitive Lie algebra is free.

Moreover, the totally primitive elements of $PQSym^{(\ell)}$ are the totally primitive elements of $PQSym$ with any coloring.

Proof

It has been done in [31] in the case of $PQSym$. But since the dendriform and codendriform structure do not involve the color alphabet C, the property is true in this case as well.

Also, colors play no role in determining whether a given element is (totally) primitive or not, so the last statement holds.

For example, the dendriform generators of PQSym have as degree generating series

$$\sum_i \mathrm{dgp}_i t^i = t + t^2 + 7t^3 + 66t^4 + 786t^5 + 11\,378t^6 + 189\,391t^7 + \cdots \tag{122}$$

so that the dendriform generators of $\mathrm{PQSym}^{(2)}$ have as degree generating series $2^i \mathrm{dgp}_i$:

$$2t + 4t^2 + 56t^3 + 1056t^4 + 25\,152t^5 + 721\,792t^6 + 24\,242\,048t^7 + \cdots \tag{123}$$

Tridendriform Structure

Conjecture 7.4

As in the case of PQSym, we conjecture that $\mathrm{PQSym}^{(\ell)}$ is a free dendriform trialgebra.

Note that there cannot be any relations, even tridendriform relations, among the elements $F_{1,c}$ where $c \in C$, so that $\mathrm{PQSym}^{(l)}$ contains the free tridendriform algebra $\mathfrak{TD}^{(\ell)}$ on ℓ generators.

Recall that, if $\mathrm{PQSym}^{(\ell)}$ is free as a tridendriform algebra, its generators have as generating series

$$\mathrm{TD} := \frac{\mathrm{PQ} - 1}{2\mathrm{PQ}^2 - \mathrm{PQ}} \tag{124}$$

where PQ is the generating series of $\mathrm{PQSym}^{(\ell)}$.

For example, the tridendriform generators of PQSym have as degree generating series

$$\sum_i \mathrm{tgp}_i t^i = t + 5t^3 + 50t^4 + 634t^5 + 9475t^6 + 163\,843t^7 + \cdots \tag{125}$$

so that the tridendriform generators of $\mathrm{PQSym}^{(2)}$ have as degree generating series $2^i \mathrm{tgp}_i$:

$$2t + 40t^3 + 800t^4 + 20\,288t^5 + 606\,400t^6 + 20\,971\,904t^7 + \cdots \tag{126}$$

Primitive Elements of $\mathrm{PQSym}^{(\ell)}$

Let $\mathscr{L}'^{(\ell)}$ be the primitive Lie algebra of $\mathrm{PQSym}^{(\ell)}$. Since Δ is not cocommutative, $\mathrm{PQSym}^{(\ell)}$ cannot be the universal enveloping algebra of $\mathscr{L}'^{(\ell)}$. But since it is cofree, it is, according to [23], the universal enveloping dipterous algebra of its primitive part $\mathscr{L}'^{(\ell)}$.

Let $G^{a,u}$ be the multiplicative basis of $\mathrm{PQSym}^{(\ell)*}$ defined by

$$\mathbf{G}^{\mathbf{a},u} = \mathbf{G}_{\mathbf{a}_1,u_1} \cdots \mathbf{G}_{\mathbf{a}_r,u_r} \tag{127}$$

where $(a,u) = (a_1,u_1) \bullet \cdots \bullet (a_1,u_r)$ is the unique maximal factorization of $(a,u) \in \mathfrak{S}_n \times C^n$ into connected colored parking functions.

Proposition 7.5

Let $V_{a,u}$ be the dual basis of $G^{a,u}$. Then, the family $\left(V_{\alpha,u}\right)_{\alpha \in} \mathcal{CP}$ is a basis of $\mathscr{L}'^{(\ell)}$. In particular, we have $\dim \mathcal{L}_n^{(\ell)} = \ell^n p_n$. Moreover, $\mathscr{L}'^{(\ell)}$ is free.

Proof

The first part of the statement follows from [4]. The second part comes from the fact that $\mathrm{PQSym}^{(\ell)}$ is bidendriform (Theorem 7.3).

For example, since $\mathscr{L}'^{(\ell)}$ is free, the generating series of the degree of its generators is (with $\ell = 2$):

$$1 - \prod_{n\geq 1}(1-t^n)^{\ell^n p_n} = 1-(1-t)^2(1-t^2)^8(1-t^3)^{88}\cdots$$
$$= 2t + 7t^2 + 72t^3 + 1276t^4 + 28\,944t^5 + 805\,288t^6 + 26\,462\,232t^7 + \cdots \quad (128)$$

and the Hilbert series of the universal enveloping algebra of $\mathrm{PQSym}^{(\ell)}$ (its domain of cocommutativity) is, again with $\ell = 2$:

$$\prod_{n\geq 1}(1-t^n)^{-\ell^n p_n} = 1 + 2t + 11t^2 + 108t^3 + 1713t^4 + 36\,470t^5 + 969\,919t^6 + 30\,847\,464t^7 + \cdots. \quad (129)$$

TYPE B ALGEBRAS

Parking Functions of Type B

In [37], Reiner defined non-crossing partitions of type B by analogy to the classical case. In our context, he defined the level 2 case. This allowed him to derive, by analogy with a simple representation theoretical result, a definition of parking functions of type B as the words on $[n]$ of size n.

We shall build another set of words, also enumerated by n^n that sheds light on the relation between type-A and type-B parking functions and provides a natural Hopf algebra structure on the latter.

First, fix two colors $0<1$. We say that a pair of words (a,u) composed of a parking function and a binary colored word is a level 2 parking function if

- The only letters of a that can have color 1 are the matches of a,
- For any letter of a of color 1, all letters equal to it and to its left also have color 1,
- All letters of a have at least once the color 0.

For example, there are 27 level 2 parking functions of size 3: there are the 16 usual ones all with full color 0, and the eleven new elements

$$(111, 100), (111, 110), (112, 100), (121, 100), (211, 010),$$
$$(113, 100), (131, 100), (311, 010), (122, 010), (212, 100), (221, 100) \quad (130)$$

The first time the first rule applies is with n=4, where one has to discard the words (1122, 0010) and (1122, 1010) since 2 is not a match of 1122. On the other hand, both words (1133, 0010) and (1133, 1010) are B_4-parking functions since 1 and 3 are matches of 1133.

Theorem 8.1

The subspace of $PQSym^{(2)}$ spanned by the F elements indexed by level 2 parking functions is a subalgebra of $PQSym^{(2)}$. The subspace of $PQSym^{(2)}$ spanned by the G elements indexed by level 2 parking functions is a subcoalgebra of $PQSym^{(2)}$.

Proof

The shifted shuffle of two **F** elements indexed by level 2 parking functions only consists in level 2 parking functions since the definition only involves matches (preserved by shifted shuffle) and positions of colors 0 and 1 on a given letter inside a word, also preserved by shifted shuffle. The second statement follows by duality.

Non-crossing Partitions of Type B

Note that in the level 1 case, non-crossing partitions are in bijection with non-decreasing parking functions. To extend this correspondence to type B, let us start with a non-decreasing parking function b (with no color). We factor it into the maximal shifted concatenation of prime non-decreasing parking functions, and we choose a color, here 0 or 1, for each factor. We obtain in this way $\binom{2\pi}{n}$ words π, which can be identified with type B non-crossing partitions.

Let

$$\mathbf{P}^{\pi} = \sum_{\mathbf{a}^{\uparrow}=\pi} \mathbf{F}_{\mathbf{a}} \tag{131}$$

where $w^{\uparrow}$ denotes the nondecreasing rearrangement of the letters of w. Then,

Theorem 8.2

The P^{π}, where π runs over the above set of non-decreasing signed parking functions, form the basis of a cocommutative Hopf subalgebra $NCPQSym^{(2)}$ of $PQSym^{(2)}$.

Proof

The subalgebra part comes from the fact that the shifted shuffle does not mix prime factors. The coalgebra part comes from the fact that deconcatenation and parkization do not mix prime factors either.

All this can be extended to higher levels in a straightforward way: allow each prime non-decreasing parking function to choose any color among ℓ and use the factorization as above. Non-decreasing parking functions are in bijection with Dyck words . Recall that these are words over a two-letter alphabet {a,b}, which can be recursively defined as follows. A Dyck word is either the empty word, or aubv, where u andv are Dyck words. When v is empty, the Dyck word is said to be irreducible. On Dyck words, the choice can be described as: each irreducible factor of a Dyck word chooses one color among ℓ . In this version, the generating series is obviously given by

$$\frac{1}{1-\ell\frac{1-\sqrt{1-4t}}{2}} \tag{132}$$

For $\ell=3$, we obtain Sequence A007854 of [40].

COLORED ANALOGS OF PLANAR BINARY TREES: $\mathrm{PBT}^{(\ell)}$

Definition of $\mathrm{PBT}^{(\ell)}$

In the case with one color, the Hopf algebra PBT of Planar binary trees initially defined by Loday and Ronco [22] can be embedded in FQSym in the following way [12] and [13]:

$$\mathbf{P}_T = \sum_{\sigma;\,\mathrm{shape}\,(\mathcal{P}(\sigma))=T} \mathbf{F}_\sigma \tag{133}$$

where $\mathcal{P}$ is a simple algorithm: it is the well-known binary search tree insertion, such as presented, for example, by Knuth in [17].

In the colored case, the definition is almost the same: a Hopf subalgebra $\mathrm{PBT}^{(\ell)}$ is spanned by the

$$\mathbf{P}_{T,u} = \sum_{(\sigma,u);\,\mathrm{shape}\,(\mathcal{P}(\sigma))=T} \mathbf{F}_{(\sigma,u)} \tag{134}$$

where u is a color word whose length is equal to the number of leaves of T.

The generating series of the dimensions of $\mathrm{PBT}^{(\ell)}$ is obviously

$$1+\ell t+2\ell^2t^2+5\ell^3t^3+14\ell^4t^4+\cdots \tag{135}$$

that is, the generating series of Catalan numbers multiplied by ℓ^n.

Algebraic Structure of $\mathrm{PBT}^{(\ell)}$

Since PBT is generated by the trees with no right branch (starting from the root), the same holds in $\mathrm{PBT}^{(\ell)}$:

Proposition 9.1

$\mathrm{PBT}^{(\ell)}$ is free over the set $P_{T,u}$, where T is a tree with no right branch.

For example, the generating series of the algebraic generators of $\mathrm{PBT}^{(\ell)}$ is

$$\ell t + \ell^2 t^2 + 2\ell^3 t^3 + 5\ell^4 t^4 + 14\ell^5 t^5 + \cdots \quad (136)$$

that is, the generating series of shifted Catalan numbers C_{n-1} multiplied by ℓ^n.

Primitive Elements of $\mathrm{PBT}^{(\ell)}$

Let $\mathcal{L}^{(\ell)}$ be the primitive Lie algebra of the algebra $\mathrm{PBT}^{(\ell)}$. Since Δ is not cocommutative, $\mathrm{PBT}^{(\ell)}$ cannot be the universal enveloping algebra of $\mathcal{L}^{(\ell)}$. But since it is cofree, it is, according to [23], the universal enveloping dipterous algebra of its primitive part $\mathcal{L}^{(\ell)}$.

Using the same arguments as in the case of $\mathrm{FQSym}^{(\ell)}$, one then proves.

Proposition 9.2

The Lie algebra $\mathcal{L}^{(\ell)}$ is free. Moreover

$$\dim \mathcal{L}_n^{(\ell)} = \ell^n C_{n-1} \quad (137)$$

For example, since $\mathcal{L}^{(\ell)}$ is free, the generating series of the degree of its generators is (with $\ell = 2$):

$$1-\prod_{n\geq 1}(1-t^n)^{\ell^n C_{n-1}} = 1-(1-t)^2(1-t^2)^4(1-t^3)^{16}(1-t^4)^{80}\dots$$
$$= 2t+3t^2+8t^3+46t^4+252t^5+1558t^6+9800t^7+\cdots \quad (138)$$

and the Hilbert series of the universal enveloping algebra of $\mathrm{PBT}^{(\ell)}$ **(its domain of cocommutativity) is, again with** ℓ **=2,**

(139)

Dendriform Structure of $\mathrm{PBT}^{(\ell)}$

Recall that PBT is the free dendriform algebra on one generator. Since colors do not play any role in determining if a given element is (totally) primitive or not, the same holds for $\mathrm{PBT}^{(\ell)}$:

Proposition 9.3

The algebra $\mathrm{PBT}^{(\ell)}$ is the free dendriform algebra on ℓ generators. It has also the structure of bidendriform bialgebra.

This provides therefore a realization of the free dendriform algebra which has been previously studied by Ronco [39].

Note 9.4: It is also possible to define a colored analog of CQSym, the Catalan Quasi-symmetric algebra, but the natural definition leads to a non-cocommutative algebra, hence not sharing the basic property of CQSym itself.

EXAMPLES

Multigraded Coinvariants and Colored Klyachko Idempotents

One of the very first applications of the theory of noncommutative symmetric functions was to provide an explanation for the following coincidence. On the one hand, consider the representation of $\mathfrak{S}_n$ in the coinvariant algebra

$$\mathcal{H}_n = \mathbb{C}[x_1, \ldots, x_n]_{\mathfrak{S}_n} = \mathbb{C}[x_1, \ldots, x_n]/\mathcal{J} \tag{140}$$

where J is the ideal generated by symmetric polynomials without constant term. It is known [24] that the graded Frobenius characteristic of the action of $\mathfrak{S}_n$ on $\mathcal{H}_n$ is

$$\begin{aligned} \mathrm{ch}_q(\mathcal{H}_n) &= \sum_k q^k \mathrm{ch}(\mathcal{H}_n)^{(k)} \\ &= (q)_n h_n\left(\frac{X}{1-q}\right) = \sum_{I \vDash n} q^{\mathrm{maj}(I)} r_I(X) \end{aligned} \tag{141}$$

where $r_I(X)$ are the ribbon Schur functions.

On the other hand, Klyachko [16] introduced a remarkable Lie idempotent in $\mathbb{C}\,\mathfrak{S}_n$

$$\kappa_n = \frac{1}{n} \sum_{\sigma \in \mathfrak{S}_n} \zeta^{\mathrm{maj}(\sigma)} \sigma \tag{142}$$

where ζ is a primitive n-th root of unity.

In terms of noncommutative symmetric functions, both expressions can be interpreted as specializations of

$$K_n(q) := (q)_n S_n\left(\frac{A}{1-q}\right) = \sum_{I \vDash n} q^{\mathrm{maj}(I)} R_I \tag{143}$$

This element can be interpreted in several ways. For example (see, e.g., [14]), it is the graded noncommutative characteristic of an action of $H_n(0)$ on coinvariants (recall that Sym_n can be identified with the Grothendieck ring $K_0(H_n(0))$). This is a projective module, hence also an S_n-module. The canonical map $e : K_0(H_0(0)) \to R(\mathfrak{S}_n) = K_0(HJ_n(1)) \simeq Sym_n$ being the commutative image, $(A=X)$, one obtains the characteristic of $C[x_1,\ldots, x_n]\mathfrak{S}_n$, which is a Hall–Littlewood function. But $K_n(q)$ can also be interpreted as an element of the descent algebra of S_n. A simple computation shows that for $q=\zeta$, a primitive n-th root of unity, $K_n(\zeta)$ is a primitive element of Sym of commutative image p_n/n, hence is a Lie idempotent (see [19]). Actually, $K_n(q)$ is the noncommutative Hall–Littlewood function $\tilde{H}_{1^n}(A:q)$, and this specialization property is a special case of a noncommutative version [11] of the classical property [11],[20] and [24].

There is a similar phenomenon here. Let $q_1,\ldots,q_n$ be independent variables, and consider the noncommutative symmetric function

$$\mathbf{K}_n(A; q_1, \ldots, q_n) := \sum_{I \vDash n} \mathbf{q}^{\mathrm{MAJ}(I)} R_I \tag{144}$$

where R_I is the ribbon basis, and

$$\mathbf{q}^{\mathrm{MAJ}(i_1,\ldots,i_r)} := (q_1 \cdots q_{i_1})^r (q_{i_1+1} \cdots q_{i_1+i_2})^{r-1} \cdots (q_{i_1+\cdots+i_{r-2}} \cdots q_{i_1+\cdots+i_{r-1}}). \tag{145}$$

For example,

$$\mathbf{K}_3 = R_3 + q_1q_2R_{21} + q_1R_{12} + q_1^2q_2R_{111} \tag{146}$$

$$\mathbf{K}_4 = R_4 + q_1q_2q_3R_{31} + q_1q_2R_{22} + q_1^2q_2^2q_3R_{211} + q_1R_{13} + q_1^2q_2q_3R_{121} + q_1^2q_2R_{112} + q_1^3q_2^2q_3R_{1111}. \tag{147}$$

Its commutative image is the multigraded characteristic of $\mathcal{H}_n$ with respect to the partition degree (cf. [3]).

One may also regard K_n as an element of $FQSym_n^{(Z)}$ by means of the identification

$$R_I(A)q_1^{\alpha_1}\cdots q_n^{\alpha_n} = \sum_{\mathrm{Des}(\sigma)=I} \mathbf{G}_{\sigma,\alpha_1\cdots\alpha_n} \tag{148}$$

and writing it as

$$\mathbf{K}_n(A; q_1, \ldots, q_n) = \sum_{I\vDash n} R_I(A)\mathbf{q}^{\mathrm{MAJ}(I)} \tag{149}$$

We shall see in the sequel that, while K_n itself is not in a generalized descent algebra, its quotient by $(1-q_1)\ (1-q_1q_2)\cdots(1-q_1\cdots q_n)$ is in fact in $MR^{(N)}$. Note that this element lives in the algebra of the extended affine Weyl group of type A

$$\widehat{\mathfrak{S}}_n = Z^n \ltimes \mathfrak{S}_n = P \ltimes \mathfrak{S}_n \tag{150}$$

where P is the weight lattice. One can also interpret it as an element of the algebra of the usual affine Weyl group

$$\widetilde{\mathfrak{S}}_n = Q \ltimes \mathfrak{S}_n \tag{151}$$

where Q is the root lattice

$$Q = \left\{ \alpha \in P \,\middle|\, \sum_{i=1}^{n} \alpha_i = 0 \right\} \tag{152}$$

This amounts to imposing the relation

$$q_1 \cdots q_n = 1 \tag{153}$$

which replaces the root of unity condition $q^n=1$ in the one-parameter case.

It has been proved by McNamara and Reutenauer [28] that under condition (153), $K_n(A;q_1\cdots,q_n)$ was a Lie idempotent in $C\tilde{\mathfrak{S}}_n$. Within the formalism of the present paper, this can be seen as follows: the authors of [28] introduce a twisted product on $\mathring{A}_n = C(x_1, \ldots, x_n)\ [\mathfrak{S}_n]$ by the formula

$$f(\mathbf{x})\sigma \cdot g(\mathbf{x})\tau = f(\mathbf{x})\sigma[g(\mathbf{x})]\sigma\tau \tag{154}$$

where permutations act on functions as automorphisms, i.e. , $\sigma(x_i) = x_{\sigma(i)}$, and in particular, on monomials by $\sigma\left[X^C\right] = X^{C\sigma-1}$. Hence,

$$(\sigma\mathbf{x}^c) \cdot (\tau\mathbf{x}^d) = \sigma\tau\mathbf{x}^{c\tau+d} \tag{155}$$

which is the same as Formula (30), so the homogeneous component of degree n of $FQSym^{(Z)}$ can be identified with a subalgebra of $\mathring{A}_n$ by setting

$$\sigma\mathbf{x}^c \equiv \mathbf{G}_\sigma\mathbf{x}^c \equiv \mathbf{G}_{\sigma,c} \tag{156}$$

McNamara and Reutenauer then introduce the formal series

$$\Theta(\mathbf{x}) = \sum_{n\geq 0}\sum_{\sigma\in\mathfrak{S}_n} \frac{\prod_{j\in \mathrm{Des}(\sigma)} x_{\sigma(1)}\cdots x_{\sigma(j)}}{\prod_{i=1}^{n}(1 - x_{\sigma(1)}\cdots x_{\sigma(i)})}\sigma \tag{157}$$

which, applying (155), and under the identification (156), can be re-written as

$$\Theta(\mathbf{x}) = \overleftarrow{\prod_{l\geq 0}}\sum_{n\geq 0} \mathbf{G}_{\mathrm{id}_n, l^n} = \overleftarrow{\prod_{l\geq 0}} \sigma_1(A^{(l)}) \tag{158}$$

Indeed, introducing the new variables

$$y_j = x_1 x_2 \cdots x_j \tag{159}$$

and applying (155) we can write

$$\Theta(\mathbf{x}) = \sum_{n\geq 0} \sum_{\sigma\in\mathfrak{S}_n} \mathbf{G}_\sigma \frac{\prod_{d\in\mathrm{Des}(\sigma)} y_d}{(1-y_1)(1-y_2)\cdots(1-y_n)} = \sum_{n\geq 0} \mathbf{K}_n(y_1,\ldots,y_{n-1}) \frac{1}{((y))_n}. \tag{160}$$

where $((y))_n = (1-y_1)(1-y_2)\cdots(1-y_n)$ and

$$\mathbf{K}_n(y_1,\ldots,y_{n-1}) = \sum_{\sigma\in\mathfrak{S}_n} \mathbf{G}_\sigma \prod_{d\in\mathrm{Des}(\sigma)} y_d \tag{161}$$

is the twisted version of the multiparameter Klyachko element introduced in [19, (11)]. By Möbius inversion over the lattice of compositions of n, we have

$$\mathbf{K}_n(y_1,\ldots,y_{n-1}) = \sum_{I\models n} R_I \cdot \prod_{d\in\mathrm{Des}(I)} y_d = \sum_{J\models n} \left(\sum_{I\geq J} (-1)^{\ell(J)-\ell(I)} \prod_{d\in\mathrm{Des}(I)} y_d \right) \tag{162}$$

and a simple argument of inclusion–exclusion shows that the coefficient of S^J in this expression is

$$\prod_{d\in\mathrm{Des}(J)} y_d \prod_{e\notin\mathrm{Des}(J)} (1-y_e) \tag{163}$$

so that

$$\mathbf{K}_n(y_1,\ldots,y_{n-1}) \frac{1}{((y))_n} = \sum_{J\models n} S^J \cdot \frac{1}{1-y_n} \prod_{d\in\mathrm{Des}(J)} \frac{y_d}{1-y_d} \tag{164}$$

which implies the expression

$$\Theta(\mathbf{x}) = \overleftarrow{\prod_{l\geq 0}} \left(\sum_{n\geq 0} S_n y_n^l \right) \tag{165}$$

Each factor of the right-hand side is grouplike (for the coproduct of $\mathbf{FQSym}^{(l)}$):

$$\Delta \sum_{n\geq 0} S_n y_n^l = \sum_{n\geq 0} \sum_{i+j=n} S_i y_i^l \otimes S_j y_j^l = \sum_{i\geq 0} S_i y_i^l \otimes \sum_{j\geq 0} S_j y_j^l \tag{166}$$

so that also

$$\Delta \Theta(\mathbf{x}) = \Theta(\mathbf{x}) \otimes \Theta(\mathbf{x}) \tag{167}$$

Extracting the term of degree n, we find

$$\Delta \mathbf{K}_n = \sum_{i+j=n} \left(\mathbf{K}_i \frac{1}{((y))_i} \otimes \mathbf{K}_j \frac{1}{((y))_j} \right) ((y))_n \tag{168}$$

so that if we send y_n to 1, all terms vanish except the extreme ones, and we get a primitive element. This is the main result of [28].

A Formula of Raney

Raney [36] gave a combinatorial interpretation of the coefficients of the unique solution $g(t) \in \mathbb{Q}[Y,Z][[t]]$ of the functional equation

$$g(t) = t \sum_{k=1}^{\ell} y_k e^{z_k g(t)} \tag{169}$$

with $g(t) = \sum_{n\geq 0} \frac{g_n}{n!} t^{n+1}$. This defining equation is of the form

$$g(t) = t\phi(g(t)) \tag{170}$$

with $\phi(u) = \sum_{k=1}^{\ell} y_k e^{z_k u}$. Hence, the coefficient g_n of t^{n+1} in $g(t)$ is

$$g_n = Y\frac{1}{n+1}[u^n]\phi(u)^{(n+1)}$$
$$= \frac{1}{n+1}\sum_{\substack{n_1+\cdots+n_\ell=n+1\\ q_1+\cdots+q_\ell=n}} \binom{n+1}{n_1,\ldots,n_\ell}\binom{n}{q_1,\ldots,q_\ell} y_1^{n_1}\cdots y_\ell^{n_\ell}(n_1z_1)^{q_1}\cdots(n_\ell z_\ell)^{q_\ell}. \tag{171}$$

Thus, $g_n \in \mathbb{N}[Y,Z]$. Its combinatorial interpretation can be mechanically derived by means of a colored version of the noncommutative Lagrange inversion formula as formulated in [32] and [8]. Consider the functional equation in $MR^{(\ell)}$

$$g = \sum_{k,n} b_k S_n^{(k)} g^n \tag{172}$$

where $S_n^{(k)} = S_n\left(A^{(k)}\right)$ is a colored complete function and b_k are noncommuting letters. We can set

$$S_n = \sum_k b_k S_n^{(k)} \tag{173}$$

so that (172) can be rewritten as

$$g = \sum_{n\geq 0} S_n g^n \tag{174}$$

and the solution of [32] reads

$$g = S^0 + S^{10} + (S^{200} + S^{110}) + \cdots$$
$$= \sum_{\pi\in \mathrm{NDPF}} S^{\mathrm{Ev}(\pi)\cdot 0}, \tag{175}$$

where NDPF is the set of nondecreasing parking functions. Note that $S^0=\sum b_k$ is a priori different from 1, and does not commute with the other S^i. Each term $S^{Ev(\pi).0}$ represents an ordered tree T in Polish notation, so that, for example

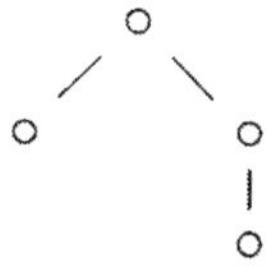

(176)

is S^{2010}.

Replacing each S^i by $\sum_{k=1}^{\ell} b_k S_i^{(k)}$, the expression $S^{Ev(\pi).0}$ becomes a sum over all ℓ-colorings of the tree T, so that one recovers the combinatorial interpretation of Raney (proved in a different way by Foata [5]): let $n=(n_1,\cdots,n_{(\ell)})$ and $q=(q_1,\cdots,q_\ell)\in N^{(\ell)}$ and let $B(n,q)$ be the set of ℓ-colored trees on $n=|n|$ vertices with n_k vertices of color k and such that the sum of the arities of vertices of color k is q_k. Then

$$g=\sum_{n\geq 0}\frac{1}{n!}\sum|B(\mathbf{n},\mathbf{q})|\mathbf{y}^{\mathbf{n}}\mathbf{z}^{\mathbf{q}} \tag{177}$$

ACKNOWLEDGMENTS

This project has been partially supported by the grant ANR-06-BLAN-0380. The authors would also like to thank the contributors of the MuPAD project, and especially of the Combinat part, for providing the development environment for their research (see [15] for an introduction to MuPAD-Combinat).

REFERENCES

1. P. Baumann, C. Hohlweg, A Solomon-type epimorphism for Mantaci–Reutenauer's algebra of a wreath product G ≀ Sn, Trans. Amer. Math. Soc. 360 (2008) 1475–1538.

2. D. Blessenohl, M. Schocker, Noncommutative Character Theory of the Symmetric Group, Imperial College Press, London, 2005, x+172 pp.
3. F. Caselli, Diagonal invariants and the refined multimahonian distribution, Preprint. arXiv:math.CO/0805.2860.
4. G. Duchamp, F. Hivert, J.-Y. Thibon, Noncommutative symmetric functions VI: free quasi-symmetric functions and related algebras, Internat. J. Algebra Comput. 12 (2002) 671–717.
5. D. Foata, La série génératrice exponentielle dans les problèmes d'énumération, Montréal Presses de l'Univ. de Montréal, 1974.
6. L. Foissy, Bidendriform bialgebras, trees, and free quasi-symmetric functions, J. Pure Appl. Algebra 209 (2) (2007) 439–459.
7. I.M. Gelfand, D. Krob, A. Lascoux, B. Leclerc, V.S. Retakh, J.-Y. Thibon, Noncommutative symmetric functions, Adv. Math. 112 (1995) 218–348.
8. I. Gessel, Noncommutative generalization and q-analog of the Lagrange inversion formula, Trans. Amer. Math. Soc. 257 (2) (1980) 455–482.
9. I. Gessel, Multipartite P-partitions and inner products of skew Schur functions, Contemp. Math. 34 (1984) 289–301. in: C. Greene (Ed.), Combinatorics and algebra..
10. I. Gessel, Enumerative applications of symmetric functions, Sém. Lothar. Combin. A 17 (1987) 17 p. (electronic).
11. F. Hivert, Hecke algebras, difference operators, and quasi-symmetric functions, Adv. Math. 155 (2) (2000) 181–238.
12. F. Hivert, J.-C. Novelli, J.-Y. Thibon, Un analogue du monoï de plaxique pour les arbres binaires de recherche, C. R. Acad. Sci. Paris Sér. I Math. 332 (2002) 577–580.
13. F. Hivert, J.-C. Novelli, J.-Y. Thibon, The algebra of binary search trees, Theoret. Comput. Sci. 339 (1) (2005) 129–165.
14. F. Hivert, J.-C. Novelli, J.-Y. Thibon, Yang–Baxter bases of 0-Hecke algebras and representation theory of 0-Ariki–Koike–Shoji algebras, Adv. Math. 205 (2006) 504–548.
15. F. Hivert, N. Thiéry, MuPAD-Combinat, an open-source package for research in algebraic combinatorics, Sém. Lothar. Combin. 51 (2004) 70 p. (electronic).
16. A.A. Klyachko, Lie elements in the tensor algebra, Sib. Math. J. 15 (1974) 1296–1304.
17. D.E. Knuth, The Art of Computer Programming, in: Sorting and Searching, vol. 3, Addison-Wesley, 1973.
18. A.G. Konheim, B. Weiss, An ocupancy discipline and applications, SIAM J. Appl. Math. 14 (1966) 1266–1274.
19. D. Krob, B. Leclerc, J.-Y. Thibon, Noncommutative symmetric functions II: transformations of alphabets, Internat. J. Algebra Comput. 7 (1997) 181–264.
20. A. Lascoux, B. Leclerc, J.-Y. Thibon, Green polynomials and Hall–Littlewood functions at roots of unity, European J. Combin. 15 (1994) 173–180.

21. J.-L. Loday, Scindement d'associativité et algèbres de Hopf, actes des journées mathématiques à la mémoire de Jean Leray, Sémin. Congr. Soc. Math. France 9 (2004) 155–172.
22. J.-L. Loday, M.O. Ronco, Hopf algebra of the planar binary trees, Adv. Math. 139 (2) (1998) 293–309.
23. J.-L. Loday, M.O. Ronco, Algèbres de Hopf colibres, C. R. Acad. Sci. Paris Sér. I Math. 337 (2003) 153–158.
24. I.G. Macdonald, Symmetric Functions and Hall Polynomials, 2nd ed., Oxford University Press, 1995.
25. C. Malvenuto, C. Reutenauer, Duality between quasi-symmetric functions and Solomon descent algebra, J. Algebra 177 (1995) 967–982.
26. R. Mantaci, C. Reutenauer, A generalization of Solomon's descent algebra for hyperoctahedral groups and wreath products, Comm. Algebra 23 (1995) 27–56.
27. P.A. McMahon, Combinatory Analysis, Cambridge University Press, 1915, 1916; Chelsea reprint, 1960.
28. P. McNamara, C. Reutenauer, P-partitions and a multi-parameter Klyachko idempotent, Electron. J. Combin. 11 (2) (2005) #R21.
29. J.-C. Novelli, J.-Y. Thibon, A Hopf algebra of parking functions, in: Proc. FPSAC'04 Conf., Vancouver.
30. J.-C. Novelli, J.-Y. Thibon, Free quasi-symmetric functions of arbitrary level, Preprint. math CO/0405597.
31. J.-C. Novelli, J.-Y. Thibon, Hopf algebras and dendriform structures arising from parking functions, Fund. Math. 193 (2007) 189–241.
32. J.-C. Novelli, J.-Y. Thibon, Noncommutative symmetric functions and Lagrange inversion, Adv. in Appl. Math. 40 (1) (2008) 8–35.
33. J.-C. Novelli, J.-Y. Thibon, A one-parameter family of dendriform identities, J. Combin. Theory Ser. A 116 (4) (2009) 864–874.
34. S. Poirier, Cycle type and descent set in wreath products, Discrete Math. 180 (1998) 315–343.
35. S. Poirier, C. Reutenauer, Algèbre de Hopf des tableaux, Ann. Sci. Math. Québec 19 (1995) 79–90.
36. G. Raney, A formal solution of $\sum_{i=j}^{\infty} A_i e^{BiX} = X$, Canad. J. Math. 16 (1964) 755–762.
37. V. Reiner, Non-crossing partitions for classical reflection groups, Preprint. Available at: ftp://s6.math.umn.edu/pub/papers/reiner/.
38. C. Reutenauer, Free Lie Algebras, Oxford University Press, 1993.
39. M.O. Ronco, Primitive elements in a free dendriform Hopf algebra, Contemp. Math. 267 (2000) 245–264.
40. N.J.A. Sloane, The on-line encyclopedia of integer sequences. http://www.research.att.com/~njas/sequences/.
41. R.P. Stanley, Parking functions and noncrossing partitions, Electron. J. Combin. 4 (1997) #2.
42. R.P. Stanley, Enumerative Combinatorics, vol. 2, Cambridge University Press, 1999.

CITATION

1. Jean-Christophe Novelli, Jean-Yves Thibon, Free quasi-symmetric functions and descent algebras for wreath products, and noncommutative multi-symmetric functions, Discrete Mathematics, Volume 310, Issue 24, 28 December 2010, Pages 3584-3606, ISSN 0012-365X, http://dx.doi.org/10.1016/j.disc.2010.09.008.

On the Complexity of Radicals in Noncommutative Rings

Chris J. Conidis[1]
Department of Mathematics, University of Chicago, Chicago IL 60637, USA

2

ABSTRACT

This article expands upon the recent work by Downey et al. (2007) [3], who classified the complexity of the nilradical and Jacobson radical of commutative rings in terms of the arithmetical hierarchy.

Let R be a computable (not necessarily commutative) ring with identity. Then it follows from the definitions that the prime radical of R is Π_1^1, and the Levitzki radical of R is Π_2^0. We show that these upper bounds for the complexity of the prime and Levitzki radicals are optimal by constructing two noncommutative computable rings with identity, such that the prime radical of one is Π_1^1-complete, while the Levitzki radical of the other is Π_2^0-complete.

INTRODUCTION

One of the first and most important questions to be studied in computable ring theory is the ideal membership problem. The analysis of this problem dates back to the work of Kronecker [8], who showed that every ideal in a computable presentation of $\mathbb{Z}[X_1, X_2,.., X_N]$ is decidable. These results were later expanded by van der Waerden [14], who showed that there does not exist a single universal splitting algorithm for factoring polynomials over all computable fields, and others. Frölich and Shepherdson [7] were first to give formal definitions in terms of recursive functions and Turing machines. They also showed, among other things, that there exists a single computable field with no splitting algorithm. By computable ring, we mean the following.

Definition 1.1

A computable ring (with identity) is a computable subset R of natural numbers, together with computable binary operations + and · on R, and elements 0, 1 ∈ R, such that (R, 0, 1,+,·) is a ring (with identity 1 ∈ R). Throughout this article we use R to denote both the domain of the ring, as well as the ordered 5-tuple (R, 0, 1,+,·).

More recently, there has been an interest in the complexity of radicals in rings in terms of the arithmetical hierarchy. In particular, Downey, Lempp, and Mileti [3] have completely classified the complexity of the nilradical and Jacobson radical in commutative computable rings, showing that the former is $\sum_1^0$-complete, while the latter is $\prod_2^0$-complete (the arithmetical and analytical hierarchies are formally introduced in the next section). We now define two radicals, which differ from the nilradical and Jacobson radical in noncommutative rings. The first is called the prime radical, while the second is known as the Levitzki radical. These radicals can be thought of as generalizations of the Jacobson radical, and some of the theorems related to the Jacobson radical can be generalized to these radicals as well. The main purpose of this article is to determine the complexity of the prime radical and Jacobson radical in a general noncommutative ring R. Let R be a (possibly noncommutative) ring with identity. By ideal we mean two-sided ideal

Definition 1.2

An ideal $P \subset R$ is prime if whenever $A\,B \subseteq P$, for ideals $A, B \subseteq R$ then either $A \subseteq P$, or else $B \subseteq P$. This is equivalent to saying that for any two elements a, b $\in$ R, we have that either a $\in$ P or b $\in$ P whenever $aRb \subseteq P$.

Definition 1.3

An ideal $P \subseteq R$ is semiprime if $A \subseteq P$ whenever A is an ideal such that $A^2 \subseteq P$. It can be shown that an ideal $P \subseteq R$ is semiprime if and only if it is an intersection of prime ideals.

Definition 1.4

The intersection of all prime ideals in R is called the prime radical of R (it is also known as the lower nilradical of R, or the Baer–McCoy radical of R). This is the smallest semiprime ideal of R. We now define the Levitzki radical of R.

Definition 1.5

A subset S of R is locally nilpotent if every subring of R (without identity) generated by a finite number of elements of S is nilpotent. It can be proved that if A and B are locally nilpotent subsets of R, then so are RAR, RBR, and A + B. From these facts it can be shown that there exists a largest locally nilpotent subset of R, and that this subset is an ideal (see Section 4).

Definition 1.6

The Levitzki radical of R is the largest locally nilpotent subset of R. Most of the typical problems that one encounters in algebra have arithmetical solutions. This means that their solutions can be expressed in relatively simple terms. For example, if R is a computable commutative ring, then by definition it follows that the nilradical of R is $\sum_1^0$, and a well-known result from classical commutative ring theory says that for every r $\in$ R, r is in the Jacobson radical of R if and only if

$$(\forall x \in R)\ (\exists a \in R) \quad \left[(1 - rx)a = 1\right]. \tag{1}$$

From this result it follows that the Jacobson radical of R is Π_2^0 (the Π comes from the $\forall$ to the far left, and the number 2 comes from the number of alternations of quantifiers in the expression). On the other hand, Downey, Lempp, and Mileti [3] have constructed computable commutative rings R0 and R1 such that the nilradical of R0 is Σ_1^0-complete, and the Jacobson radical of R_1 is Π_2^0-complete, thus showing that the simplest characterization of the nilradical is the standard definition, while the simplest characterization of the Jacobson radical is (1) above. Many more examples of arithmetical ring-theoretic constructions exist, see for example [2,3,5,6].

Above the arithmetical hierarchy lies the analytic hierarchy. Analytical sets are more complex than arithmetical sets, because to define an analytic set one is allowed to quantify over both number variables (as in the arithmetical case), as well as function (or set) variables. The reader should note that every arithmetical set is analytical, but not vice versa. For example, the standard definition of the Jacobson radical of a commutative ring R is the intersection of all maximal ideals in R. Since this definition quantifies over all the maximal ideals of R, it follows from the definition that the Jacobson radical of a computable ring is analytic. However, (1) above gives a different (arithmetical) characterization of the Jacobson radical, from which it follows that the Jacobson radical of a computable ring is always in fact arithmetical. In the next section we define a well-known set called WF (the set of computable indices for well-founded trees) that is analytic but not arithmetic.

When a set $X \subseteq \omega$ is shown to be analytical but not arithmetical, it implies that function or set quantifiers are necessary to define X via a computable predicate. For example, in Section 3, we construct a computable ring R whose prime radical is Π_1^1-complete. It follows that the prime radical of R is analytical but not arithmetical. One consequence of this construction is that any effective definition of the prime radical must involve quantifying over sets of natural numbers. In other words, one must say something like "the prime radical of a ring R is the intersection of all the prime ideals in R" (here we are quantifying over all prime ideals of R). The superscript 1 in Π^1_1 says that we are allowed to quantify over sets, while the subscript 1 says that only one set quantifier is necessary in the definition of the prime radical.

By definition, it follows that if R is a computable ring, then the prime radical of R is a Π^1_1 set, and the Levitzki radical of R is a Π^0_2 set. The main purpose of this article is to show that these upper bounds on the complexity of the prime radical and Levitzki radical are sharp, by constructing computable rings R0 and R1 such that the prime radical of R0 is Π^1_1-complete, and the Levitzki radical of R1 is Π^0_2-complete. More formally, the main goal of this article is to prove Theorems 1.7 and 1.8 below. The proof of Theorem 1.7 is given in Section 3, while the proof of Theorem 1.8 is given in Section 4. The formal definition of completeness is given in the next section, but, intuitively, to say that a set X is Γ -complete means that:

1. X belongs to the complexity class Γ .
2. The complexity of X is maximal among Γ -sets, in the sense that every Γ -set can be (computably) reduced to X.

Our main goal in this article is to prove the following theorems.

Theorem 1.7

There exists a noncommutative computable ring R such that the prime radical of R is Π1 1 -complete.

Theorem 1.8

There exists a noncommutative computable ring R such that the Levitzki radical of R is Π0 2 -complete.

PRELIMINARIES

Background

Let ω denote the set of natural numbers, i.e. ω = {0, 1, 2, 3,...}. By ring we mean a (possibly noncommutative) ring with identity. We assume that the reader is familiar with the basic definitions of ring theory, as well as those of (oracle) Turing machines and (relative) computation. Standard texts in commutative ring theory include [1, 4, 10, 11]. A standard text on noncommutative rings is [9]. Two standard references in computability theory are [12, 13].

Fix a computable bijection $p_2 : \omega \times \omega \to \omega$, and numbers x, y ∈ ω. We will denote $p_2(x, y)$ by x, y . Furthermore, for every n ∈ ω, n≥3, define a function pn $:\omega^n \to \omega$ by

$$p_n(x_0, x_1, x_2, \ldots, x_{n-1}) = \langle x_0, p_{n-1}(x_1, x_2, \ldots, x_{n-1})\rangle.$$

It follows (by induction) that p_n is a computable bijection, and that

$$p_n(x_0, x_1, x_2, \ldots, x_{n-1}) = \langle x_0, \langle x_1, \langle x_2, \langle \ldots \langle x_{n-2}, x_{n-1}\rangle\rangle \ldots \rangle.$$

For every n, $x_0, x_1, \ldots, x_{n-1}$ ∈ ω, we let

$$\langle x_0, x_1, x_2, \ldots, x_{n-1}\rangle = p_n(x_0, x_1, x_2, \ldots, x_{n-1}).$$

We now review the construction of the arithmetical hierarchy. Fix natural numbers m,n≥1.

1. We say that a set $X \subseteq \omega^m$ is $\sum_n^0$, and write $X \in \sum_n^0$, if there exists a computable set $A \subseteq \omega^{n+m}$ such that for every $x_1, x_2, \ldots, x_m \in \omega$ we have that

   $$(x_1, x_2, \ldots, x_m) \in X \quad \Leftrightarrow \quad \exists a_1 \, \forall a_2 \, \exists \ldots \, Q a_n \quad [(x_1, x_2, \ldots, x_m, a_1, a_2, \ldots, a_n) \in A],$$

   where Q is ∃ if n is odd, and ∀ if n is even.

2. A set $X \subseteq \omega^m$ is $\prod_n^0$, and write $X \in \prod_n^0$, if there exists a computable set $A \subseteq \omega^{n+m}$ such that for every $x_1, x_2, \ldots, x_m \in \omega$ we have that

$$(x_1, x_2, \ldots, x_m) \in X \quad \Leftrightarrow \quad \forall a_1 \, \exists a_2 \, \forall \ldots \, Q a_n \quad [(x_1, x_2, \ldots, x_m, a_1, a_2, \ldots, a_n) \in A],$$

where Q is ∃ if n is even, and ∀ if n is odd.

Definition 2.1

A $\sum_n^0$ (resp $\prod_n^0$) set X ⊆ ω is called $\sum_n^0$ ($\prod_n^0$)-complete if for every set Y ∈ $\sum_n^0$ ($\prod_n^0$) there is a computable function hY :ω → ω such that for every n ∈ ω, n ∈ Y if and only if h_Y (n) ∈ X.

For our purposes, we are most interested in $\prod_2^0$ sets, since the proof of Theorem 1.8 involves reducing every $\prod_2^0$ set to the Levitzki radical of a noncommutative computable ring. With this in mind, we state the following standard computability-theoretic result. Recall that if $\{\varphi_e\}_{e\in\omega}$ is an effective listing of the partial computable functions, then, for every e ∈ ω, the eth computably enumerable (c.e.) set is defined to be

$$W_e = \{x \in \omega \colon \phi_e(x)\downarrow\}.$$

Proposition 2.2

The set

$$\text{Inf} = \{e \in \omega \colon |W_e| = \infty\}$$

is $\prod_2^0$-complete.

Therefore, to show that a given set X is $\prod_2^0$-complete, it suffices to find a computable function h such that for all n ∈ ω, n ∈ Inf if and only if h(n) ∈ X. We now define what it means for a set X ⊂ ω to be $\prod_1^1$. Recall that ω^ω denotes the set of functions f :ω → ω.

Definition 2.3

We say that a set $X \subset \omega^m$ is Π_1^1, and write $X \in \Pi_1^1$, if there exists a number $n \in \omega$, and a computable set $A \subseteq \omega^\omega \times \omega^{m+n}$, such that for all x_1, $x_2, \ldots, x_m \in \omega$ we have that

$$(x_1, x_2, \ldots, x_m) \in X \quad \Leftrightarrow \quad \forall f\ \exists a_1\ \forall a_2\ \ldots\ Q a_n\ \ \big[(f, a_1, a_2, \ldots, a_n, x_1, x_2, \ldots, x_m) \in A\big].$$

where Q is $\forall$ if n is even, and $\exists$ if n is odd.

A well-known result says that, without loss of generality, we can always assume that $n = 1$ in Definition 2.3.

Definition 2.4

A Π_1^1 set $X \subseteq \omega$ is called Π_1^1-complete if for every set $Y \in \Pi_1^1$, there is a computable function $h_Y : \omega \to \omega$ such that for every $n \in \omega$, $n \in Y$ if and only if $h_Y(n) \in X$.

We now construct an example of a Π_1^1-complete set called WF (the set of computable indices for well-founded trees).

Let $\omega^{<\omega}$ denote the set of strings of natural numbers. For any $\sigma, \tau \in \omega^{<\omega}$ write $\sigma \subseteq \tau$ to mean that σ is an initial segment of τ. A nonempty subset T of $\omega^{<\omega}$ is closed downwards if for every $\sigma \in T$ and every $\tau \in \omega^{<\omega}$ such that $\tau \subseteq \sigma$, we have that $\tau \in T$. We call subsets of $\omega^{<\omega}$ that are closed downwards trees.

Let $T \subseteq \omega^{<\omega}$ be a tree, and $\sigma \in T$. We say that σ is an extendible node if there exists an infinite path through T extending σ – i.e. if there exists $f \in \omega^\omega$ such that for every $n \in \omega$, $f\,1\,n \in T$. Here $f\,1\,n = \langle\ f(0), f(1) \ldots, f(n-1) \in \omega^{<\omega}$ denotes the first n bits of f. We also say that T is well founded if no $\sigma \in T$ is an extendible node. By definition it follows that if T is a computable tree, then the property of T being well-founded is Π_1^1. It turns out that this property is also Π_1^1-complete.

Proposition 2.5

Let $\{T_e\}_{e\in\omega}$ be an effective listing of all computable trees. Then the set

$$\mathrm{WF} = \{e \in \omega\colon\ T_e \text{ is a well-founded tree}\}$$

is Π^1_1-complete

Hence, to show that a given set X is Π^1_1-complete, it suffices to find a computable function h such that for all $n \in \omega$, $n \in \mathrm{WF}$ if and only if $h(n) \in X$.

Now that we have given the reader the necessary preliminaries, we are ready to prove Theorems 1.7 and 1.8. Throughout this article, R will always denote a (possibly) noncommutative ring with identity. In Section 3 we prove Theorem 1.7, and in Section 4 we prove Theorem 1.8. As an aside, it may interest the reader to know that in a general noncommutative ring R, if B denotes the prime radical of R, L denotes the Levitzki radical of R, N denotes the nilradical of R, and J denotes the Jacobson radical of R, then we have that

$$B \subseteq L \subseteq N \subseteq J,$$

and the inclusions are strict in general.

PRIME RADICAL

Recall that the prime radical of a (possibly) noncommutative ring R is defined to be the intersection of all the prime ideals of R. From this it follows that the prime radical of a computable (possibly) noncommutative ring R is a Π^1_1 set. Hence, the most that one could hope for is to construct a computable (noncommutative) ring R whose prime radical is Π^1_1-complete. With this observation in mind, we prove the following theorem.

Theorem 1.7

There exists a noncommutative computable ring R such that the prime radical of R is Π_1^1-complete

First, however, we require some definitions. Let R be a ring.

Definition 3.1

For any elements a, b ∈ R, we say that a divides b if b is contained in the (two-sided) ideal generated by a, i.e. b ∈ a .

Definition 3.2

A nonempty set S ⊆ R is called an m-system if, for any a, b ∈ S, there exists r ∈ R such that arb ∈ S.

Definition 3.3

Let R be a ring with identity. For any two-sided ideal I ⊆ R, define

$$\sqrt{I} = \{s \in R:\ \text{every } m\text{-system containing } s \text{ meets } I\}.$$

Theorem 3.4

The prime radical of R is equal to $\sqrt{0}$.

Let $\mathbb{Q}[\vec{X}] = \mathbb{Q}[X_0, X_1, X_2,...]$ be the noncommutative polynomial ring in countably many indeterminates over the field of rational numbers. $\mathbb{Q}$ Throughout the remainder of this section we will only consider rings R of the form R = $\mathbb{Q}[\vec{X}]/I$, for some two-sided ideal I ⊆ $\mathbb{Q}[\vec{X}]$. In this case we use the notation X ∈ R to denote the image of X ∈ $\mathbb{Q}[\vec{X}]$ under the canonical map $\varphi : \mathbb{Q}[\vec{X}] \to R$. By monomial, we mean nonconstant monomial. An element r ∈ R is said to be a monomial if it is equivalent to the image of a monomial under φ.

Definition 3.5

A nonempty set $S \subseteq R$ is a monomial m-system if, for any $a, b \in S$, there is a monomial $r \in R$ such that $arb \in S$.

We now prove a simple proposition that allows us to construct monomial m-systems in R.

Proposition 3.6

Let $x_0 = \overline{X}_n \in R$, for some $n \in \omega$, and for every $i > 0$, let $x_i = x_{i-1}m_{i-1}x_{i-1}$ for some monomial $m_{i-1} \in R$. Then, if $i, j \in \omega$ are given, with $i \leq j$, there exist monomials $m_0, m_1 \in R$ such that $x_{j+1} = x_i m_0 x_j$ and $x_{j+1} = x_j m_1 x_i$. It follows that the set $X = \{x_0, x_1, x_2, \ldots\} \subset R$ is a monomial m-system.

Proof

We prove the existence of m_0. The proof of the existence of m_1 is similar

The proof is by induction on $j = \max\{i, j\}$. If $j = 0$, then since i j, we have that $i = j = 0$ and by definition of $x_1 = x_0 m_0 x_0$, the proposition holds. A similar argument shows that the proposition holds if $i = j$, so assume that $i < j$. Before we prove the induction step, we make the

obvious observation that, by construction, for every $n \in \omega$, $x_n \in \mathbb{Q}[\vec{X}]$ is a monomial.

If $j > 0$, assume that the proposition holds for $j - 1$; we shall show that the proposition also

holds for j. By the induction hypothesis and the fact that $i < j$, there is a monomial m

such that $x_j = x_i m' x_{j=1}$. Now, we have that $x_{j+1} = x_j m_j x_j$, and so $x_{j+i} = x_i(m' x_{j-1} m_j) x_j$ Hence, the desired monomial mo is equal to $m' \text{-} x_{j-1}$mj.This proves the induction step, and thus completes the proof of the proposition.

Having given the necessary background, we are now ready to prove Theorem 1.7.

Proof of Theorem 1.7

Let $\mathbb{Q}\,[\overrightarrow{X\omega^{<\omega}}]$ be the polynomial ring over the field of rational numbers $\mathbb{Q}$, with

indeterminates X_σ, for every $\sigma \in \omega^{<\omega}$. Let $T \subset \omega^{<\omega}$ be a computable tree containing every node of length 1, and such that the set of extendible nodes in T of length 1 is Π^1_1-complete. Such a tree $T \subset \omega^{<\omega}$ may be constructed as follows. First, put all nodes of length 1 in T . Then, if $\{T_e\}_{e\in\omega}$ is an effective listing of the computable trees in $\omega^{<\omega}$, for every $e \in \omega$ put the tree T_e above the node $\langle e \rangle$ (of length 1) into T . By the construction of T , it follows that T is a computable tree in $\omega^{<\omega}$.

We shall construct a computable ring R of the form R = ¤$[\overrightarrow{X\omega^{<\omega}}]$ /I, for some (computable) ideal I ⊂ ¤$[\overrightarrow{X\omega^{<\omega}}]$ such that I is generated by a computable set of monomials. Furthermore, the prime radical of R shall be П1 1 -complete.

Let a computable function F $:\omega^{<\omega} \to$ ¤$[\overrightarrow{X\omega^{<\omega}}]$ be defined as follows. F $(\emptyset) = 1$, and if $\sigma \in \omega^{<\omega}$ is such that $|\sigma| > 0$, then define F (σ) = F $(\sigma -)X_\sigma$ F (σ^-), where σ^- is the unique initial segment of σ such that $|\sigma^-|=|\sigma|$ - 1. Note that (by induction we have that) for every node $\rho \in \omega<\omega$, F (ρ) is a monomial of degree $2^{|\rho|} - 1$, unless $\rho = \emptyset$ in which case F $(\rho) = 1$. Using the function F , we now construct the computable ideal I such that R =¤$[\overrightarrow{X\omega^{<\omega}}]$.

Let I ⊆ ¤$[\overrightarrow{X\omega^{<\omega}}]$ be the ideal generated by the monomials m ∈ ¤$[\overrightarrow{X\omega^{<\omega}}]$ such that m does not divide any monomial of the form F (σ), $\sigma \in T$.

Note that if a monomial m ∈ ¤$[\overrightarrow{X\omega^{<\omega}}]$ contains an occurrence of some indeterminate X_σ, where $\sigma \in/ T$, then it follows that m cannot divide any element of the form F (τ), $\tau \in T$, and thus by definition of I we have that m ∈ I. We also have the following proposition.

Proposition 3.7

Let $m \in$ ¤$[\overrightarrow{X\omega^{<\omega}}]$ be a monomial, and let $\sigma \in \omega^{<\omega}$ be maximal such that X_σ appears in m. Then $m \notin I$ if and only if m divides $F(\sigma)$.

Proof

If m divides $F(\sigma)$, then by definition of I it follows that $m \notin I$.

Now, suppose that $m \notin I$. Then there is some $\tau \in \omega^{<\omega}$ such that m divides $F(\tau)$. Note that we must have $\sigma \subseteq \tau$ since otherwise, by the construction of F, we know that X_σ does not appear in $F(\tau)$, and so m cannot divide $F(\tau)$. It suffices to show that if $\tau \supsetneq \sigma$, then m divides $F(\tau^-)$, where τ^- is the unique initial segment of τ of length $|\tau| - 1$. Suppose that $\tau \supsetneq \sigma$. By definition of F, we have that $F(\tau) = F(\tau^-) X_\tau F(\tau^-)$. Now, by definition of σ and the fact that $\tau \supsetneq \sigma$, we know that the indeterminate X_τ does not appear in m. Therefore, since m divides $F(\tau) = F(\tau^-) X_\tau F(\tau^-)$, it follows that m must also divide $F(\tau^-)$.

Corollary 3.8

The ideal $I \subset R$ is computable.

Proof

Since the ideal I is generated by monomials, it follows that a polynomial $p \in \mathbb{Q}[\overrightarrow{X\omega^{<\omega}}]$ is in the ideal I if and only if every monomial summand m of p is in I. Proposition 3.7 gives a method for deciding whether or not a given monomial is in I, and so it also gives a method for deciding whether or not $p \in I$.

The following corollary is a consequence of the proof of Proposition 3.7.

Corollary 3.9

If $m \in$ ¤$[\overrightarrow{X\omega^{<\omega}}]$ is a monomial such that $m \in/ I$, and if $\sigma \in \omega^{<\omega}$ is maximal such that X_σ appears in m, then X_σ is unique. In other words, if σ, τ $\in \omega^{<\omega}$ and X_σ and X_τ appear in m, then σ and τ are comparable.

Now that we have constructed the computable ring $R =$ ¤$[\overrightarrow{X\omega^{<\omega}}]$ $/I$, it remains to show that $\sqrt{\langle 0\rangle} \subseteq R$ is Π_1^1-complete. With this in mind, we prove the following proposition.

Proposition 3.10

For every $\sigma \in \omega^{<\omega}$, $X_\sigma \in/ \sqrt{\langle 0\rangle} \subseteq R$ if and only if there is an infinite path through T extending $\sigma \in T$.

Proof

First, we claim that if $\sigma \in \omega^{<\omega}$ is an extendible node of T , then there is a monomial m-system containing X_σ but not containing 0. The proof is as follows. Let $f \in \omega^\omega$ be an infinite path through T extending σ . Then, by Proposition 3.6, and the constructions of F and I, it follows that the image of F restricted to f (in the quotient R) is a monomial m-system containing X_σ but not containing 0.

Now, let $\sigma \in T \subset \omega^{<\omega}$, and suppose that there is an m-system S in R containing X_σ , but not containing 0. In this case we claim that there is an infinite path in T extending σ . To construct such a path, first set $y_0 = X_\sigma \in$ ¤$[\overrightarrow{X\omega^{<\omega}}]$, and for every number $n > 0$, let yn $\in$¤$[\overrightarrow{X\omega^{<\omega}}]$, $y_n \in/ I$, be of the form $y_n = y_{n-1} r_{n-1} y_{n-1}$, for some $r_{n-1} \in$ ¤$[\overrightarrow{X\omega^{<\omega}}]$.

To prove that there is an infinite path in T extending $\sigma \in T$, we first prove the following lemma which says that there is an infinite, finitely branching tree $T_0 \subseteq T$ above σ . Then we apply König's Lemma to the tree $T_0 \subseteq \omega^{<\omega}$ to get an infinite path in T0 $\subseteq$ T extending σ.

Lemma 3.11

There is an infinite, finitely branching tree $T_0 \subseteq T$ such that for every $\tau \subseteq \sigma$, $\tau \in T_0$, and for all $\tau \in T_0$, if $\tau \not\subset \sigma$, then $\tau \supset \sigma$.

Proof

We begin by giving several definitions and constructions which shall aid us in the proof of

Lemma 3.11. Let $m \in \mathbb{Q}[\overrightarrow{X\omega^{<\omega}}]$ be a monomial and $p \in \mathbb{Q}[\overrightarrow{X\omega^{<\omega}}]$ be a polynomial.

Definition 3.12

We say that m is an essential monomial summand of p if m is a summand of p such

that $m \in/ I$ (i.e. $\bar{m} = 0 \in R$).

For every $n \in \omega$, define

$Y_n = \tau \in \omega^{<\omega}$: X_τ appears in an essential monomial summand of $y_n\}$

Now, define $T_0 \subseteq \omega^{<\omega}$ to be the downward closure of the set

$$\{\rho \in \omega^{<\omega}: (\exists n \in \omega)\,[\rho \in Y_n]\},$$

and for every $s \in \omega$, let T_0^s be the downward closure of the set

$$\{\rho \in \omega^{<\omega}: (\exists n \leqslant s)\,[\rho \in Y_n]\}.$$

By definition, it follows that $T_0 = \cup_{s \in \omega} T_0^s$ and T_0 is a tree. Also, since for every $n \in \omega$, the set of $\sigma \in \omega^{<\omega}$ such that X_σ appears in yn $\in \mathbb{Q}[\overrightarrow{X\omega^{<\omega}}]$ is finite, it follows that for every $s \in \omega$, T_0^s is a finite (and hence finitely branching) tree. Moreover, recall that if $\tau \in/ T$ then (by definition of I) it follows that $X_\tau \in I$. Therefore, if $\tau \in \omega^{<\omega}$ is such that $\tau \in/ T$ and X_τ appears in some monomial summand m of y_n for some $n \in \omega$, then m is not an essential monomial summand of y_n. Hence, by definition of Y_n, $n \in \omega$, and T_0, it follows that T_0 is a subtree of T . It remains to be shown that every initial segment of σ is in T_0, every node $\tau \in T_0$ is comparable to σ, and that T_0 is an infinite, finitely branching tree.

By assumption, we know that $\sigma \in T$. It follows that $y_0 = X_\sigma \in/ I$, and thus X_σ is an essential summand of y_0. Therefore, by the construction of T_0, it follows that every initial segment of σ belongs to T_0. Furthermore, by induction on $n \in \omega$, it follows that for every $n \in \omega$ and every monomial summand m of y_n, X_σ appears in m. Now, by Corollary 3.9 and the definition of T_0, it follows that if $\tau \in T_0$ then τ is comparable to σ.

We now show that T_0 is infinite by showing that T_0 contains nodes of arbitrarily large length. First note that (by induction on $n \in \omega$ it follows that) for all $n \in \omega$, every monomial summand of y_n has degree at least 2^n. Furthermore, by definition of I, it follows that if $m \in \mathbb{Q}[\overrightarrow{X\omega^{<\omega}}]$. is a monomial of degree 2_n, then m cannot divide F (ρ) for any $\rho \in T$ of length less than n (since in this case F (ρ) has degree $2^{|\rho|} -1 < 2^{|\rho|}$). Hence, by definition of I, if m is an essential summand of y_n, then m must divide some F (ρ), where $|\rho|$ n. Now, by Proposition 3.7, it follows that if m is an essential summand of y_n, then m contains an occurrence of some indeterminate $X\rho$, $\rho \in T$, $|\rho|$ n. We have now shown that every essential monomial summand m of y_n contains an occurrence of some indeterminate X_ρ , where $\rho \in T$ and $|\rho|$ n. By assumption, we have that $y_n \in/ I$, for every $n \in \omega$. Hence, for every $n \in \omega$ there exists an essential monomial summand m of y_n. Therefore, by definition of T_0, it follows that T_0 contains nodes of arbitrarily large length. Next, we complete the proof of Lemma 3.11 by showing that T_0 is a finitely branching tree.

To show that T_0 is finitely branching, fix a node $\tau \in T_0$, and let $n \in \omega$ be large enough so that every essential monomial summand m of y_n contains an occurrence of an indeterminate of the form X_ρ , for some node $\rho \in \omega^{<\omega}$ such that $|\rho| > |\tau|$ (the previous paragraph explains why such an n exists).

We claim that the sets

$$S_0 = \left\{\rho \in T_0 \colon |\rho| = |\tau| + 1 \text{ and } \rho \supset \tau\right\}$$

and

$$S_1 = \left\{\rho \in T_0^n \colon |\rho| = |\tau| + 1 \text{ and } \rho \supset \tau\right\}$$

are equal. Since $T_0 = \bigcup_{s\in\omega} T_0^s$, it follows that $S_0 \supseteq S_1$. We need to show that $S_0 \subseteq S_1$. Suppose, for a contradiction, that there exists a node $\rho \in S_0 \setminus S_1$. Then, by definition of S_0, S_1, T_0, T n 0, it follows that there exists a number $m > n$ and a node $\rho_0 \supseteq \rho$ such that $X_{\rho 0}$ appears in an essential monomial summand of y_m. However, by definition of n, and the fact that $m > n$, it follows that every monomial summand of y_m contains an occurrence of some indeterminate X_λ, where $|\lambda| > |\tau|$ and $\lambda(|\tau| + 1) \in S_1$. Since $\rho \notin S_1$, it follows that λ and ρ_0 are incomparable nodes for all such λ. Now, by Corollary 3.9 it follows that no monomial summand of y_m in which $X_{\rho 0}$ appears is an essential monomial summand of y_m, a contradiction. Thus, we have shown that $S_0 = S_1$, and therefore T0 is finitely branching.

Applying König's Lemma to the infinite, finitely branching tree $T_0 \subseteq T \subset \omega^{<\omega}$ yields an infinite path $f \in \omega^\omega$ through T extending $\sigma \in T$. This completes the proof of Proposition 3.10.

We now construct a computable function h: $\omega \to R$ by setting, for every number $n \in \omega$ corresponding to the node $\langle n\rangle \in \omega^{<\omega}$, $h(n) = \vec{X}_{\langle n\rangle} \in R$. Proposition 3.10 says that $n \in WF$ if and only if $h(n) \in \sqrt{\langle 0\rangle} \subset R$. Therefore, $\sqrt{\langle 0\rangle} \subset R$ is Π1 1 -complete. This completes the proof of Theorem 1.7.

LEVITZKI RADICAL

In this section we prove the following theorem.

Theorem 1.8

There exists a noncommutative computable ring R such that the Levitzki radical of R is Π_2^0-complete.

First, however, we require two definitions and a proposition. Let R be a ring.

Definition 4.1

A set $S \subset R$ is locally nilpotent if, for any finite subset $S_0 = \{s_0, s_1, \ldots, s_n\} \subseteq S$, there is a number $N = N(S_0) \in \omega$ such that any product of N elements from $\{s_0, s_1, \ldots, s_n\}$ is zero. This is equivalent to Definition 1.5.

The following proposition is standard. We omit its proof.

Proposition 4.2

Let I, J be locally nilpotent one-sided ideals in R. Then R I R, and I + J are locally nilpotent.

We now define the Levitzki radical of a ring R

Definition 4.3

The Levitzki radical of R, $L \subset R$, is the largest locally nilpotent ideal in R. By Proposition 4.2, we have that

$$L = \{x \in R : xR \text{ is locally nilpotent}\},$$

and that $L \subset R$ is an ideal.

We are now ready to prove Theorem 1.8.

Proof of Theorem 1.8

We have already remarked that, for all $x \in R$, $x \in L$ if and only if xR is locally nilpotent. In other words, $x \in L$ if and only if

$$\left(\forall \langle x_0, x_1, \ldots, x_n \rangle \in xR\right) (\exists N) \left(\forall \sigma \in n^N\right) \left[\prod_{i<|\sigma|} x_{\sigma(i)} = 0\right].$$

Hence, it follows that if R is a computable ring, then $L \in \Pi^0_2$ (the last quantifier in the expression above is bounded[2]). We now show that this (upper) bound on the complexity of the Levitzki radical is sharp by constructing a computable ring R whose Levitzki radical is Π^0_2-complete. The rest of this section is dedicated to the construction of R.

R shall be a quotient of the form $\mathbb{Q}[\vec{X}]/I$, where $\mathbb{Q}[\vec{X}] = \mathbb{Q}[X_0, X_1, X_2, \ldots]$ is the noncommutative polynomial ring in countably many variables over the rational numbers $\mathbb{Q}$, and $I \subset \mathbb{Q}[\vec{X}]$ is a (twosided) ideal. We construct $\cup_{s\in\omega}$ in stages such that for all $s \in \omega$, $I_s \subseteq I_{s+1}$.

At stage 0 define I0 to be the computable ideal in $\mathbb{Q}[\vec{X}]$generated by the monomials $m \in \mathbb{Q}[\vec{X}]$such that there are numbers $e_0, e_1, i, j \in \omega$ with $e_0 \neq e_1$, and indeterminates $X_{(e0,i)}$, $X_{(e1),j}$ both occurring in m.

At stage s + 1, we are given the computable ideal I_s, and add to it all monomials m of degree greater than s + 1, such that the only indeterminates appearing in m are in the set $\{X_{(e,0)}, X_{(e,1)}, \ldots, X_{(e,s)}\}$, where $e \in \omega$ is such that $W_{e,s+1} \neq W_{e,s}$ (without loss of generality we assume that at every stage s there exists a unique $e \in \omega$ such that $W_{e,s+1} \neq W_{e,s}$).

We now verify that $I = \cup_{s\in\omega} I_s$ is computable. To see why this is the case, first note that I is generated by monomials. Thus, it suffices to show that the set of monomials that generate I, $M \subset I$, is a computable set. To see why this is the case, note that, by the construction of $I = \cup_{s\in\omega} I_s$, we have that for any monomial $m \in \mathbb{Q}[\vec{X}]$of degree d, $m \in I$ if and only if m ∈ Id. For every X ∈¤[$\vec{X}$], let X denote the image of X under the canonical quotient map φ: ℚ [X] → R.

Recall that the set Inf = {e ∈ ω: W_e is infinite} is Π_2^0-complete. Therefore, to show that L ⊂ R is Π_2^0-complete, it suffices to exhibit a computable function h: ω → ℚ [$\vec{X}$] such that for every e ∈ ω, e ∈ Inf if and only if h (e) ∈ L. We claim that the computable map h: ω → R such that h(e) = X⟨e,0⟩ e,0 satisfies this condition.

To verify that the function h above has the desired property, we shall prove that for every e ∈ ω, X⟨e,0⟩ ∈ L if and only if We is infinite. It suffices to prove the following proposition.

Proposition 4.4

For every e ∈ ω, the right ideal $\overline{X_{\langle e,0\rangle}}$ ·R, is locally nilpotent if and only if We is infinite.

Proof

First, suppose that W_e is finite. Then there is a stage s_e ∈ ω such that for all s≥s_e we have $W_{e,s+1}$ = W_e . Fix a number n ∈ ω. By the construction of R = ℚ [$\vec{X}$]/I, we have that X = (X⟨e,0⟩ · $X_{(e,se+1)}$)^n ∈/ I_{s_e} , since, by the construction of I_{s_e} , we have that Xe,0 ∈/ I_{s_e} , and no (monomial) generator of I_{s_e} contains an appearance of the indeterminate $X_{\langle e,s_e+1\rangle}$ (and $X_{(e,s_e+1)}$ appears in X). Furthermore, since at all stages t se we do not enumerate any new elements into W_e, then by the construction of I = ∪ $_{s∈N}$ I_s , it follows that we do not enumerate X into I at any stage t≥se.

Therefore, X ∉ I, and so $\overline{X} \neq 0 \in R$. It follows that $\overline{X_{\langle e,0\rangle}}$ is not locally nilpotent, and hence $\overline{X_{\langle e,0\rangle}} \notin L$.

Now suppose that W_e is infinite. Let $\overline{m}_0, \overline{m}_1, \ldots, \overline{m}_\Pi \in R$ be nonzero, and let $M \in \omega$ be large so that, for all $0 \leq i \leq n$ and $X_{\langle e,0 \rangle}$ occurring in mi, we have that $M > \max\{e, j\}$. We shall show that there exists a number $N \in \omega$ such that

$$\prod_{k=0}^{N} (\overline{X_{\langle e,0 \rangle}}) \cdot \overline{m_{i_k}} = 0, \tag{2}$$

where $i_k \in \{0, 1, \ldots, n\}$ for all $0 \leq k \leq N$. Without loss of generality, we can assume that for all $0 \leq i \leq n$, we have that $m_i \in ¤[\vec{X}]$ is a monomial. Therefore, it follows that the product in (2) above is a monomial of degree greater than or equal to N. Now, by the construction of R, if we let $N = s$, where $s \in \omega$ is the least stage greater than M such that $W_{e,s+1} \neq W_{e,s}$ (note that s exists, since We is infinite), then we have that (2) holds, as required. Hence,$\cdot X_{\langle e,0 \rangle}$ R is locally nilpotent.

This completes the proof of Theorem 1.8.

REFERENCES

1. M.F. Atiyah, I.G. MacDonald, Introduction to Commutative Algebra, Perseus, 1969.
2. C.J. Conidis, Chain conditions in computable rings, Trans. Amer. Math. Soc., in press.
3. R.G. Downey, S. Lempp, J.R. Mileti, Ideals in computable rings, J. Algebra 314 (2007) 872–887.
4. D.S. Dummit, R.M. Foote, Abstract Algebra, John Wiley & Sons, 1999.
5. H.M. Friedman, S.G. Simpson, R.L. Smith, Countable algebra and set existence axioms, Ann. Pure Appl. Logic 25 (1983) 141–181.
6. H.M. Friedman, S.G. Simpson, R.L. Smith, Addendum to: "Countable algebra and set existence axioms", Ann. Pure Appl. Logic 28 (1985) 319–320.
7. A. Frölich, J.C. Shepherdson, Effective procedures in field theory, Philos. Trans. R. Soc. Lond. Ser. A Math. Phys. Eng. Sci. 248 (1956) 407–432.
8. L. Kronecker, Grundzüge einer arithmetischen Theorie der algebraischen Groössen, J. Reine Angew. Math. 92 (1882) 1–122.

9. T.Y. Lam, A First Course in Noncommutative Rings, second ed., Springer-Verlag, 2001.
10. S. Lang, Algebra, Springer-Verlag, 1993.
11. H. Matsumura, Commutative Ring Theory, Cambridge University Press, 2004.
12. H. Rogers, Theory of Recursive Functions and Effective Computability, MIT Press, 1987.
13. R.I. Soare, Recursively Enumerable Sets and Degrees, Springer-Verlag, 1987.
14. B.L. van der Waerden, Algebra, vol. 1, Springer-Verlag, 2003.

CITATION

1. Chris J. Conidis, On the complexity of radicals in noncommutative rings, Journal of Algebra, Volume 322, Issue 10, 15 November 2009, Pages 3670-3680, ISSN 0021-8693, http://dx.doi.org/10.1016/j.jalgebra.2009.07.039.

Cohomology of Tails, Tate–Vogel Cohomology, and Noncommutative Serre Duality over Koszul Quiver Algebras

3

Roberto Martinez Villa[a] *and*
Alex Martsinkovsky[b]
[a]Instituto de Mathemáticas, UNAM, Apdo Postal 61-B, Morelia, Mich., Mexico, 58089
[b]Mathematics Department, Northeastern University, Boston, MA 02115, USA

ABSTRACT

The main result of the paper shows that, under Koszul duality between quiver algebras, cohomology of tails is identified with graded Vogel cohomology. As an application, a new proof of the noncommutative Serre duality over generalized Artin–Schelter regular Koszul quiver algebras is given. It is deduced from a similar formula over an arbitrary (i.e., not necessarily Koszul) Frobenius algebra, which turns out to be equivalent to the Auslander–Reiten formula. As another application, it is shown that, over a generalized Artin–Schelter regular Koszul quiver algebra, any algebra automorphism appearing in the noncommutative Serre duality formula is closely related, under Koszul duality, to the Nakayama automorphism of the Koszul-dual algebra.

INTRODUCTION

The goal of this paper is to show that, for Koszul quiver algebras, cohomology of tails and Vogel cohomology correspond to each other under Koszul duality. More precisely, Koszul duality gives rise to a functorial isomorphism between tail cohomology of Koszul modules and Vogel cohomology of the Koszul-dual modules. As an application, we give a new proof of the noncommutative Serre duality for tails over a generalized Artin–Schelter regular Koszul algebra. By passing to the Koszul-dual algebra, we show that the Serre duality is a consequence of the duality between the functors Ext and Tor and the availability of the dimension shift to degree zero in Tate cohomology of the Koszul-dual algebra, which is necessarily Frobenius. As a by-product, we obtained a new duality formula for finite modules over a Frobenius algebra. This formula should be viewed as an analog of Serre duality for Frobenius algebras. As another application, we show that, over a generalized Artin–Schelter regular Koszul algebra, any automorphism appearing in the Serre duality formula is closely related, under Koszul duality, to the Nakayama automorphism of the Koszul-dual algebra.

Here is a brief outline of the paper. In Section 2 we set up the notation and recall basic facts about quiver algebras and their categories of graded modules.

In Section 3 we recall the definition and basic properties of Koszul quiver algebras.

In Section 4 we stabilize the category of Koszul modules over a Koszul quiver algebra by the graded Jacobson radical J , which can be viewed as an endofunctor, and show that the result is naturally isomorphic to the stabilization1 of the Koszul-dual category by the syzygy operation Ω. A specific choice of Ω, made in the previous section, makes it into a functor. We then pass to the Koszul-dual category modulo projectives. This passage does not, in general, give rise to an isomorphism of the directed systems of Hom-sets, but one of the main results of the paper, Theorem 4, asserts that the isomorphism is restored in the limit.

In Section 5, after recalling basic definitions, we interpret the above result as a functorial isomorphism between tail cohomology of Koszul modules over a Koszul quiver algebra and graded Vogel cohomology of the Koszul-dual modules over the Koszul-dual algebra. Assuming the Noetherian property, together with finite global dimension, this result is extended to arbitrary finitely generated modules. In particular, this applies to the classical case of cohomology of coherent sheaves on a projective space.

In Section 6 we give a streamlined account of graded Tate cohomology over a selfinjective ring. Assuming that the contravariant argument is finitely presented and using the (graded) transpose, one immediately identifies Tate cohomology in nonpositive cohomological degrees as the appropriate values of the Tor-functor.

In Section 7 we deal with modules of type FP_∞ over generalized Artin–Schelter regular Koszul algebras and the Serre duality formula for such modules, established earlier by the first author [9]. The Koszul-dual algebra is selfinjective and, being basic, is necessarily Frobenius. We show that the Koszul-dual of the twist of a Koszul module by the algebra automorphism appearing in the Serre duality formula is isomorphic, modulo projectives, to the twist of the Koszul-dual module by the inverse of the Nakayama automorphism of the Koszul-dual algebra.

In Section 8, we reprove, as another application of the main results, the just mentioned Serre duality formula for modules of type FP_∞ over generalized Artin–Schelter regular Koszul algebras. To the best of our knowledge, a noncommutative generalization of Serre duality was first given by A.B. Verëvkin ca. 1989 (see [13] and [14] for further details). For other approaches, see [16] and [8]. The proof given in this paper uses properties of Tate cohomology over selfinjective rings and is new even for coherent sheaves on a projective space. Besides providing a simple and clear approach to Serre duality, it also "explains" the seemingly strange fact that the dual of a cohomology class is still a cohomology, rather than homology, class. The explanation is clear once we pass, via Koszul duality, to the dual algebra, which is necessarily Frobenius. Under this passage, cohomology of tails transforms into Tate cohomology, and it is a well-known fact that Tate cohomology can be expressed as Tate homology. In the end, this werewolf-like behavior of Tate cohomology is explained by applying dimension shift and Auslander's transpose. Moreover the duality formula over the Frobenius algebra holds in general, i.e., without assuming the Koszul property. It is also equivalent to the Auslander–Reiten formula.

NOTATION AND PRELIMINARIES

Let be a field and Q a finite quiver. The associated path algebra $\mathbb{k}Q$ is considered graded in such a way that the vertices are of degree zero and arrows are of degree one. A quotient of $\mathbb{k}Q$ will be called a quiver algebra. Unless stated otherwise, we shall work with algebras of the form $\Lambda = \mathbb{k}Q/I$, where I is a homogeneous ideal contained in the square of the ideal generated by the arrows of Q. Thus our algebras Λ will be nonnegatively graded and Λ_0 will be the product of a finite number of copies of the field, one copy for each vertex of Q. In particular, Λ_0 will be a finite-dimensional -algebra. Writing Λ as $\coprod_{i\geq 0} \Lambda_i$, the two-sided ideal $J := \coprod_{i\geq 1} \Lambda_i$ is called the graded (Jacobson) radical of Λ.

All modules in this paper will be left modules, unless stated otherwise. Also, all Λ-modules

will be $\mathbb{Z}$ -graded, i.e., $M := \coprod_{i\in\mathbb{Z}} M_i$ and $\Lambda_i M_j \subset M_{i+j}$ for all integers i and j .In particular, each M_i will be a Λ_0-bimodule and, therefore, a -vector space. We visualize such graded modules as vertical strips with the increasing order of the degrees going down. Thus elements of smaller degrees are positioned higher than elements of larger degrees. The seemingly redundant phrase "M is generated in the highest degree n" means, in particular, that all generators of M are in degree n. Since Λ is nonnegatively graded, this implies that n is the smallest integer such that $M_n \neq 0$. Thus the term highest refers, perhaps confusingly for some readers, to the visual image of the module and not to the usual ordering of the integers. The symbol$M_{\geq s}$, respectively, $M_{\leq s}$, denotes the direct sum $\coprod_{i\geq s} M_i$, respectively, $\coprod_{i\leq s} M_i$ of Λ_0-bimodules. Notice that the former is a Λ-submodule of M. If M is generated in the highest degree n, then $M_{\geq s} = J^{S-n}M$ for all integers $s \geq n$.

The shifts M[k], $k \in \mathbb{Z}$ of M are defined as follows: $M[k]_j := M_{k+j}$. Visually, the choice of a positive (respectively, negative) k results in an upward (respectively, downward) displacement. The support of M is defined as the set of all indices i such that $M_i \neq 0$.

If M and N are graded Λ-modules, a degree n map is a Λ-map f: M → N such that $f(M_i) \subset N_{i+n}$ for all $i \in \mathbb{Z}$. For each integer k, the symbol f [k] will denote the map M[k] → N[k] which sends x ∈ M[k] to f (x). Notice that f [k] and f have the same degree n. The Λ_0-bimodule of all degree n maps from M to N will be denoted $\mathrm{Hom}_\Lambda(M,N)_n$. It is easy to check that there are canonical isomorphisms:

$$\mathrm{Hom}_\Lambda\big(M, N[n]\big)_0 \cong \mathrm{Hom}_\Lambda(M, N)_n \cong \mathrm{Hom}_\Lambda\big(M[-n], N\big)_0$$

The graded Λ_0-bimodule $\coprod_n \mathrm{Hom}_\Lambda(M,N)_n$ will be denoted $\mathrm{Homgr}_\Lambda(M,N)$. If M or N is finitely generated, then $\mathrm{Hom}_\Lambda(M,N) \cong \mathrm{Homgr}_\Lambda(M,N)$. All graded (left) Λ-modules and degree zero maps between them form an abelian category, denoted GrModΛ. Thus $\mathrm{Hom}_{\mathrm{GrMod}\Lambda}(M, N) = \mathrm{Hom}_\Lambda(M,N)_0$. The notation Grmod$\Lambda$ will be reserved for the category of finitely generated graded Λ-modules. Unless stated otherwise, module homomorphisms in this paper will be assumed to be of degree zero.

Recall that a graded Λ-module M is said to be locally finite if, for each integer n, the Λ_0-module M_n is of finite length. Since Λ_0 is finite-dimensional over , this is equivalent to saying that each M_n is a finite-dimensional -vector space. The full subcategory of GrModΛ consisting of all locally finite (left) Λ-modules is an abelian category, denoted LFGrModΛ. The duality $D(-) := \mathrm{Hom}_{\mathbb{k}}(-,\mathbb{k})$ on the category of finite-dimensional k-vector spaces extends to a duality, denoted by the same latter,

$$D : \mathrm{LFGr\,Mod}\,\Lambda \to \mathrm{LFGr\,Mod}\,\Lambda^{\mathrm{op}}$$

given by $D(M)_j := D(M_{-j})$ for all $j \in \mathbb{Z}$. In our notation, $D(-) := \mathrm{Homgr}_{\mathbb{k}}(-,\mathbb{k})$.

We shall also make use of the functor $\underline{\mathrm{Hom}}_{\mathrm{GrMod}\Lambda}(-,-)$, colloquially referred to as "Hom modulo projectives". More precisely, a degree n map f :M →N is said to factor through a projective if there exist an integer i , a projective module P, and maps g :M →P of degree n−i and h :P →N of degree i such that f = hg. Let $P(M,N)_p$ be the maps in $\mathrm{Hom}(M,N)_p$ which factor through projectives. Then $\underline{\mathrm{Hom}}_{\Lambda}(M,N)_p$ is defined as $\mathrm{Hom}_{\Lambda}(M,N)_p / P(M,N)$ and $\underline{\mathrm{Hom}}_{\mathrm{GrMod}\Lambda}(M,N)$ is defined as $\underline{\mathrm{Hom}}_{\Lambda}(M,N)_0$. We also define $\underline{\mathrm{Hom}}\mathrm{gr}_{\Lambda}(M,N)$ as $\coprod_p \underline{\mathrm{Hom}}_{\Lambda}(M,N)_p$.

For graded Λ-modules M and N, the Λ_0-bimodule $\mathrm{Ext}^k_{\Lambda}(M,N)_n$, can be defined as either the kth right-derived functor of $\mathrm{Hom}_{\mathrm{GrMod}\Lambda}(M,N[n]) = \mathrm{Hom}_{\Lambda}(M,N)_n$ or as the quotient of the Λ_0-bimodule $\mathrm{Hom}_{\mathrm{GrMod}\Lambda}(\Omega^k M,N)$ by the sub-bimodule of the boundaries. Here Ω^k is the k-fold composition of $\Omega = \Omega^1$ with itself, Ω being the kernel of a graded projective precover. By definition, graded projective precovers will always be of degree zero.Whenever graded projective covers are available, Ω will be understood as the kernel of a graded projective cover. By [5, Proposition 2.4, 5], finitely generated graded modules over graded quiver algebras have graded projective covers. Notice that, in our notation, $\mathrm{Ext}^k_{\mathrm{GrMod}\Lambda}(M,N) = \mathrm{Ext}^k_{\Lambda}(M,N)_0$. Similarly, one defines $\mathrm{Tor}_k^{\mathrm{GrMod}\Lambda}(M,N)$ as the degree zero part of $\mathrm{Tor}_k^{\Lambda}(M,N)$. Here M is a Λ^{op}-module.

Recall that a module is said to be of type FP_∞ if it has a projective resolution consisting of finitely generated projectives. If M or N is of type FP_∞, then $\mathrm{Ext}^k_{\Lambda}(M,N) \cong \coprod_n \mathrm{Ext}^k_{\Lambda}(M,N)_n$. In that case, the Λ_0-bimodule $\coprod_k \mathrm{Ext}^k_{\Lambda}(M,N) \cong \coprod_k \coprod_n \mathrm{Ext}^k_{\Lambda}(M,N)_n$ is bigraded. The grading by the cohomological degree k will be referred to as the external grading, and the grading by the integer n as the internal grading. To distinguish, when necessary, the two gradings we shall use the symbol [i] to denote the degree i external shift, and the symbol (i) to denote the degree i internal shift. Notice however that we shall always consider the external grading as primary, in the sense that degree zero maps between Ext-modules, whenever they arise, will be understood as degree zero maps with respect to the external grading.

It is easy to check that the dimension shift respects the internal grading. In other words, there are natural isomorphisms of Λ_0-bimodules $\mathrm{Ext}^k_\Lambda(M,N) \cong \mathrm{Ext}^{k-1}_\Lambda(\Omega M,N)_n$ for each k≥2 and for each n. It is also easy to see that, for all integers k and n, there are canonical isomorphisms of Λ_0-bimodules

$$\mathrm{Ext}^k_\Lambda(M, N[n])_0 \cong \mathrm{Ext}^k_\Lambda(M, N)_n \cong \mathrm{Ext}^k_\Lambda(M[-n], N)_0$$

Finally, it is not difficult to check that, for $\theta \in \mathrm{Ext}^1_\Lambda(M,N)_n$, one can always realize ϑ by an extension $0 \to N \xrightarrow{f} L \xrightarrow{g} M \to 0$, where L is a graded Λ-module, the maps f and g are homogeneous, and deg f + deg g =−n. Moreover, one of the two degrees can be chosen arbitrarily.

We end this section with two observations concerning graded projectives.

Lemma 1: Let Λ = Q/I be a graded quiver algebra and P a finitely generated graded projective Λ-module. Then P is a direct sum of graded projectives generated in highest degrees.

Proof: The images e1, . . . , en of the length zero paths in Q form a complete set of primitive orthogonal idempotents of Λ. By the assumption, there are degree zero maps

$$P \to \coprod \Lambda[m_i] \to P$$

whose composition is the identity. Reducing these maps modulo the graded radical J and using the fact that Λ/J is semisimple, we have that P/J P is isomorphic to a direct sum $\coprod S_j[m_j]$ of shifted graded simples. Since P →P/J P is a projective cover, P is a projective cover of $\coprod S_j[m_j]$.

But for each j the projective cover of S_j is of the form Λ_{enj} for a suitable n_j. Thus P is isomorphic to $\coprod \Lambda e_{n_j}[m_j]$.

Lemma 2: Let $\Lambda = kQ/I$ be a graded quiver algebra, M and N Λ-modules finitely generated in the same highest degree l, and $f \in \mathrm{Hom}_\Lambda(M,N)_0$. If f is nonzero and factors through a graded projective P, then M has a nonzero projective summand.

Proof: Without loss of generality, we may assume that P is a projective cover of N and is thus generated in the highest degree l. Let J be the graded radical of Λ. Writing f = gh with $h \in \mathrm{Hom}_\Lambda(M,P)_0$ and $g \in \mathrm{Hom}_\Lambda(P,N)_0$ we have, since f is nonzero and since M is generated in the highest degree l, that $f_l = g_l h_l : M/JM \to N/JN$ is nonzero, too. Hence $h_l : M/JM \to P/JP$ is nonzero and, therefore, since P/JP is semisimple, the image of hl is the direct sum of graded simples. Then the direct sum of the corresponding projective covers is a direct summand of M.

KOSZUL QUIVER ALGEBRAS AND KOSZUL DUALITY

In this section we shall review some basic facts about Koszul algebras which will be needed later. All missing proofs can be found in [5] and [6].

Definition 1: Let $\Lambda := kQ/I$ be a graded quiver algebra with graded radical J, and M a graded Λ-module of type FP_∞ with graded minimal projective resolution

$$\cdots \to P_i \to P_{i-1} \to \cdots \to P_0 \to M \to 0.$$

a) M is said to be Koszul if each nonzero Pi is generated in degree i.
b) Λ is said to be Koszul if all graded simple Λ-modules are Koszul.

Given modules L, M, and N over an algebra Λ and classes $h \in \mathrm{Ext}^k_\Lambda(L,M)$ and $g \in \mathrm{Ext}^l_\Lambda(M,N)$ one defines the product $gh \in \mathrm{Ext}^{k+l}_\Lambda(L,N)$ by composing representatives of h and g in the sense of Yoneda. Thus if g and h are thought of as classes of chain maps between shifted projective resolutions, then gh is obtained by applying a representative of g after a representative of h. This multiplication makes $\coprod_k \operatorname{Ext}^k_\Lambda(L, M)$ into a left graded module over the (cohomologically) graded k-algebra $\coprod_k \operatorname{Ext}^k_\Lambda(M, M)$.

The following theorem describes some of the main properties of Koszul algebras (see [6, Theorem 2.2 and Proposition 5].

Theorem 1: Let $\Lambda := kQ/I$ be a Koszul algebra and $\Gamma := \coprod_{k\geq 0} \operatorname{Ext}^k_\Lambda(\Lambda_0, \Lambda_0)$ its Yoneda algebra. Let $F : \operatorname{GrMod} \Lambda \to \operatorname{GrMod} \Gamma$ be the functor 3 given by $F(-) := \coprod_{k\geq 0} \operatorname{Ext}^k_\Lambda(-, \Lambda_0)$. Then:

(i) If M is a finitely generated graded semisimple Λ-module generated in degree zero, then F(M) is a finitely generated graded projective Γ-module generated in degree zero.

(ii) If M is a finitely generated graded projective Λ-module generated in degree zero, then F(M) is a finitely generated graded semisimple Γ-module generated in degree zero.

(iii) Λ is quadratic, i.e., the ideal I is generated by k-linear combinations of paths of length two.

(iv) The Yoneda algebra $\Gamma = \coprod_{k\geq 0} \operatorname{Ext}^k_\Lambda(\Lambda_0, \Lambda_0)$ is Koszul.

(v) The algebra Γ is isomorphic to kQ^{op}/L, where Q^{op} is the opposite quiver of Q and L is the ideal generated by its degree two part L_2, which is obtained as follows.

Let V_2 (respectively, V_2^o) be the vector subspace of kQ (respectively, kQ^{op}) spanned by the paths of length two and $\langle -,- \rangle : V2 \times V_2^o \to k$ be the bilinear form defined on the basis elements by

$$\langle \alpha\beta, \beta'\alpha' \rangle := \begin{cases} 1 & \text{if } \alpha = \alpha' \text{ and } \beta = \beta', \\ 0 & \text{otherwise.} \end{cases}$$

Then $L_2 := I_2^{\perp}$, where $I_2 := I \cap V_2$.

(vi) The opposite algebra of Λ is Koszul.

(vii) If M is a Koszul Λ-module, then both $J^kM[k]$ and $\Omega^kM[k]$ are Koszul for each $k \geq 1$.

(viii) Let K_Λ and K_Γ denote the categories of Koszul modules over Λ and, respectively, Γ with degree zero maps. Then the functor F: GrModΛ→GrModΓ defined above induces a duality, denoted by the same letter, $F : K_\Lambda \to K_\Gamma$ such that, for any Koszul Λ-module M, there are isomorphisms

$F(J_\Lambda^k M[k]) \simeq \Omega^k F(M)[k]$ and $F(\Omega^k M[k]) \simeq J_\Gamma^k F(M)[k]$ for each

$k \geq 1$. The functor $G: K_\Gamma \to K_\Lambda$, given by $G(-) = \coprod_{k \geq 0} \mathrm{Ext}_\Gamma^k(-, \Gamma_0)$, is a quasi-inverse of F.

We now briefly outline the construction of the isomorphisms in part (viii) (see [6, Proposition 5.1.3-5] for more details). To obtain the first isomorphism, apply $\mathrm{Hom}_\Lambda(-,\Lambda/J)$ to the short exact sequence $0 \to JM \to M \to M/JM \to 0$. The resulting long cohomology exact sequence will break into short exact sequences

0→Extn−1Λ (JM, Λ/J) →ExtnΛ(M/JM,Λ/J)→ExtnΛ(M,Λ/J)→0,

which yield the exact sequence

$0 \to F(JM)[-1] \to F(M/JM) \to F(M) \to 0.$

Since F(M/JM)→F(M) is a projective cover, one has an isomorphism F(JM[1]) F(JM)(−1) F(JM) ΩF(M)[1]. Notice that the two middle modules differ from each other only in the internal grading and, therefore, are isomorphic, as Γ -modules, in the external grading. The general statement now follows by induction.

Remarks

(1) A priori, the symbol$\Omega F(M)$ does not naturally depend on M, as the syzygy operation is only defined up to projective equivalence. Sometimes, however, a specific choice of Ω on modules and on maps does result in a functor. Inspired by the cohomological construction of the previous paragraph, we shall now make such a choice of Ω on Koszul Γ-modules by defining $\Omega F(M)$ as $F(J\,M[1])[-1]$. As a result, we have an actual equality $\Omega F(M)[1] = F(J\,M[1])$ rather than an isomorphism. Henceforth, this equality will always be assumed. This definition extends in an obvious way to shifts of Koszul modules and we have, by induction, that $\Omega nF(M) = F(J\,nM[n])[-n]$.

(2) Suppose $f : M \to N$ is a degree zero homomorphismbetween Koszul modules. Utilizingthenaturalityoflongcohomologyexactsequences,appliedtotheprevious construction, we have a natural map $F(Jf[1])[-1] : F(JN[1])[-1] \to F(JM[1])[-1]$. The same construction shows that this map is a lifting $\Omega F(N)\to\Omega F(M)$ of $F(f)$, where $\Omega F(N)$ (respectively,$\Omega F(M)$), according to the previous remark, is understood as $F(J\,N[1])[-1]$ (respectively, $F(J\,M[1])[-1]$). We now define $\Omega(F(f)) := F(J\,f[1])[-1]$. Thus the action of Ω on Koszul modules and maps between them is defined via long cohomology exact sequences. As a result, Ω becomes a functor on the category of (shifts of) Koszul Γ-modules. Moreover, using the fact that Λ is the Koszul-dual of Γ, and making similar choices for the syzygy operation on Koszul Λ-modules and morphisms between them, we have that Ω is a functor on the category of (shifts of) Koszul Λ-modules.

(3) For any graded Λ-module M and for any integer i, the graded Γ-modules F(M) and $F(M[i]) \cong F(M)(-i)$ are canonically and naturally isomorphic.

(4) Part (vii) of Theorem 1 shows that $M_{\geq k}[k]$ is Koszul for any Koszul Λ-module M and any integer k. With the choices made in the first two remarks, we now have

Lemma 3: The isomorphism $F(J^k_\Lambda M[k]) \to \Omega^k F(M)[k]$ is natural in M for each $k \geq 1$. To construct the second isomorphism in Theorem 1(viii), one applies the functor $\mathrm{Hom}_\Lambda(-,\Lambda/J)$ to the exact sequence $0 \to \Omega M \to P \to M \to 0$ and obtains isomorphisms $\mathrm{Ext}^{n-1}_\Lambda(\Omega M, \Lambda/J) \simeq \mathrm{Ext}^{n-1}_\Lambda(M, \Lambda/J)$ for each $n \geq 1$. An induction argument gives the desired result. Since the established isomorphisms are parts of a long cohomology exact sequence, we have

Lemma 4: The isomorphism $F(\Omega^k M[k]) \to J^k_\Gamma F(M)[k]$ is natural in M for each $k \geq 1$. The next result easily follows from parts (vii) and (viii) of Theorem 1.

Lemma 5: Let $\Lambda := kQ/I$ be a Koszul algebra with Yoneda algebra Γ and $F : \mathrm{GrMod}\Lambda \to \mathrm{GrMod}\Gamma$ the functor described in the previous theorem. If M and N are Koszul Λ-modules, then, for each integer k, we have natural isomorphisms:

$$\mathrm{Hom}_{\mathrm{GrMod}\,\Lambda}(J^k M, J^k N) \cong \mathrm{Hom}_{\mathrm{GrMod}\,\Gamma}(F(J^k N), F(J^k M))$$
$$\cong \mathrm{Hom}_{\mathrm{GrMod}\,\Gamma}(\Omega^k F(N), \Omega^k F(M)).$$

Proof: Notice first that $\mathrm{Hom}_{\mathrm{GrMod}\,\Lambda}(J^k M, J^k N) \cong \mathrm{Hom}_{\mathrm{GrMod}\,\Lambda}(J^k M[k], J^k N[k])$. The latter is naturally isomorphic to $\mathrm{Hom}_{\mathrm{GrMod}\,\Gamma}(F(J^k N[k]), F(J^k M[k]))$ by Koszul duality. The first claim now follows from remark (3). The naturality of the second isomorphism follows from Lemma 3.

Lemma 6: Under the assumptions of the previous lemma, for each integer k, there is a natural isomorphism:

$$\mathrm{Ext}^1_{\mathrm{Gr\,Mod}\,\Lambda}(M[k], N[k]) \cong \mathrm{Ext}^1_{\mathrm{Gr\,Mod}\,\Gamma}(F(N[k]), F(M[k]))$$

Proof: Choose a representative $0 \to N[k] \to L \to M[k] \to 0$, with maps of degree zero, of $\theta \in \mathrm{Ext}^1_{\mathrm{GrMod}\Lambda}(M[k], N[k])$. As the end modules are generated in degree k, the same is true for L. The Horseshoe Lemma shows that L is a degree k shift of a Koszul module. The induced sequence $0 \to N[k]/JN[k] \to L/JL \to M[k]/JM[k] \to 0$ of semisimple modules is exact, being given by the restrictions of the maps in ϑ to the common highest degree –k. Therefore, the direct sum of the projective covers of the end modules is a projective cover of the middle module. Thus the same is true for the original sequence. Hence, starting with minimal projective resolutions of N[k] and M[k], the Horseshoe Lemma yields a minimal projective resolution of L. In other words, the original short exact sequence induces a short exact sequence of the minimal projective resolutions. Applying the functor $\mathrm{Hom}_\Lambda(-, \Lambda_0)$ and passing to the corresponding long cohomology exact sequencewe have, because of the minimality of the projective resolution of L, that the connecting homomorphism is zero. This gives a short exact sequence $0 \to F(M[k]) \to F(L) \to F(N[k]) \to 0$, which we take as the image of θ in $\mathrm{Ext}^1_{\mathrm{GrMod}\Gamma}(F(N[k]),F(M[k]))$. The naturality of the defined map is obvious. Finally, the quasi-inverse G of F gives rise, via the same construction, to the inverse of that map.

The next result shows that, under mild assumptions, graded modules of finite projective dimension can be approximated by Koszul modules. For a proof, see [9, Proposition 2.4].

Theorem 2: Let $\Lambda := kQ/I$ be a Koszul algebra and M a graded Λ-module of finite projective dimension. Assume that M is of type FP_∞. Then there exists an integer k such that the shift $M_{\geq k}[k]$ of the truncated module $M_{\geq k}$ is Koszul.

ITERATING J AND Ω

Let Γ be a Koszul algebra with graded radical J and Yoneda algebra Λ. In this section we want to show that, with appropriate shifts, the stabilization of $\mathrm{Hom}_{\mathrm{GrMod}\Gamma}(-,-)$ by J is naturally isomorphic, under Koszul duality, to the stabilization of $\mathrm{Hom}_{\mathrm{GrMod}\Lambda}(-,-)$ by Ω. In Section 5, we shall interpret this result as a statement that Koszul duality induces an isomorphism between cohomology of tails over Γ and stable cohomology over Λ. We begin with simple observations.

Lemma 7: Let Γ be a nonnegatively graded ring and M and N graded Γ-modules with disjoint supports.4 Then $\mathrm{Hom}_{\mathrm{GrMod}\Gamma}(M,N) = 0$. If every element of the support of N is less than every element of the support of M, then $\mathrm{Ext}^1_{\mathrm{GrMod}\Gamma}(M,N) = 0$.

Proof: The first claim is immediate. Now let a short exact sequence $0 \to N \xrightarrow{f} X \xrightarrow{g} M \to 0$ with deg f = deg g = 0 be a representative of an element of $\mathrm{Ext}^1_{\mathrm{GrMod}\Gamma}(M,N)$ and let s − 1 be the largest element in the support of N. Then the assumption on the supports implies that the image of f coincides with $X_{\leq s-1}$. Therefore $X_{\leq s-1}$ is a Γ-submodule of X. Since the ring is concentrated in the nonnegative degrees, $X_{\geq s}$ is also a Γ-submodule of X, and we have a direct sum decomposition $X = X_{\leq s-1} \coprod X_{\geq s}$. This shows that f is a split embedding.

Lemma 8: Let Γ be a nonnegatively graded ring with graded Jacobson radical J, and M and N graded Γ-modules generated, respectively, in degrees s and t, $s \geq t$. Then the canonical inclusion $j : J^{s-t}N \mapsto N$ induces natural isomorphisms

$$\mathrm{Hom}_{\mathrm{Gr\,Mod}\,\Gamma}(M, j) : \mathrm{Hom}_{\mathrm{Gr\,Mod}\,\Gamma}(M, J^{s-t}N) \to \mathrm{Hom}_{\mathrm{Gr\,Mod}\,\Gamma}(M, N)$$

$$\mathrm{Ext}^1_{\mathrm{Gr\,Mod}\,\Gamma}(M, j) : \mathrm{Ext}^1_{\mathrm{Gr\,Mod}\,\Gamma}(M, J^{s-t}N) \to \mathrm{Ext}^1_{\mathrm{Gr\,Mod}\,\Gamma}(M, N)$$

Proof: The naturality is obvious. Since N is generated in the highest degree t, $J^{s-t}N = N_{\geq s}$ and, therefore, the support of $N / J^{s-t}N$ is above the support of M. We can now finish the proof by applying the functor $\mathrm{Hom}_{\mathrm{GrMod}\Gamma}(M,-)$ to the short exact sequence $0 \to J^{s-t}N \to N \to N / J^{s-t}N \to 0$ and using the previous lemma.

Remark: As the proof of the lemma shows, the requirement that M be generated in the highest degree s can be weakened: it suffices to assume that the support of M is in the degrees bigger than or equal to s.

Theorem 3: Let Γ be a Koszul algebra with graded radical J and Yoneda algebra Λ and let $F : \mathrm{mod}\Gamma \to \mathrm{mod}\Lambda$ be the functor $F(-) = \coprod_{i \geq 0} \mathrm{Ext}^i_{\Gamma}(-, \Gamma / J)$. Then, for any pair of Koszul Γ-modules M and N, any nonnegative integer n, any integer p, and any integer $k \geq \max\{-p, 0\}$ there is a natural, in M and N, isomorphism of directed over k systems:

$$\mathrm{Ext}^n_{\mathrm{Gr\,Mod}\,\Gamma}(J^k M, N[p]) \cong \mathrm{Ext}^n_{\mathrm{Gr\,Mod}\,\Lambda}(\Omega^{k+p} F(N)[p], \Omega^k F(M)).$$

As a consequence, there is a natural isomorphism:

$$\mathrm{Ext}^n_{\mathrm{Gr\,Mod}\,\Gamma}(J^k M, N[p]) \cong \varinjlim_{k \geqslant \max\{-p,0\}} \mathrm{Ext}^n_{\mathrm{Gr\,Mod}\,\Lambda}(\Omega^{k+p} F(N)[p], \Omega^k F(M))$$

Proof: We shall consider separately the cases n = 0 and n > 0. Assume first that n = 0. By Lemma 8, $\mathrm{Hom}_{\mathrm{GrMod}\Gamma}(J^{k}M, N[p])$ is naturally isomorphic to $\mathrm{Hom}_{\mathrm{GrMod}\Gamma}(J^{k}M, J^{k+p}N[p])$, which, by Lemma 5, is naturally isomorphic to $\mathrm{Hom}_{\mathrm{GrMod}\Lambda}(F(J^{k+p}N[p]), F(J^{k}M))$. Now we want to construct an explicit natural isomorphism to

$$\mathrm{Hom}_{\mathrm{Gr\,Mod}\,\Lambda}\left(\Omega^{k+p}F(N)[p], \Omega^{k}F(M)\right)$$

This will be done with the help of the construction in Remarks on p. 64. Let $f : J^{k}M \to J^{k+p}N[p]$ be an arbitrary degree zero map. Then

$$f[k] \in \mathrm{Hom}_{\mathrm{Gr\,Mod}\,\Gamma}\left(J^{k}M[k], J^{k+p}N[k+p]\right)$$

is also an arbitrary map. By Koszul duality, F(f [k]) is an arbitrary element of

$$\mathrm{Hom}_{\mathrm{Gr\,Mod}\,\Lambda}\left(F\left(J^{k+p}N[k+p]\right), F\left(J^{k}M[k]\right)\right)$$

and, therefore, F(f) is an arbitrary element of $\mathrm{Hom}_{\mathrm{GrMod}\Lambda}(F(J^{k+p}N[p]), F(J^{k}M))$. By our construction of the syzygy functor, $F(J^{k+p}N[k+p]) = \Omega^{k+p}F(N)[k+p]$ and $F(J^{k}M[k]) = \Omega^{k}F(M)[k]$. We now take $F(f[k])[-k] \in \mathrm{Hom}_{\mathrm{GrMod}\Lambda}(\Omega^{k}+pF(N)[p], \Omega^{k}F(M))$ as the image of F(f). It is immediate that the constructed map is a natural isomorphism. This gives the desired assertion for n = 0.

Nowassumethatn≥1.ByLemma8anddimensionshift, $\mathrm{Ext}^{n}_{\mathrm{GrMod}\Gamma}(J^{k}M, N[p])$ is naturally isomorphic to $\mathrm{Ext}^{1}_{\mathrm{GrMod}\Gamma}(\Omega^{n-1}J^{k}M, J^{k+n-1+p}N[p])$. The latter is naturally isomorphic to $\mathrm{Ext}^{1}_{\mathrm{GrMod}\Lambda}(F(J^{k+n-1+p}N[p]), F(\Omega^{n-1}J^{k}M))$ by Theorem 1(vii) on p. 63 and Lemma 6 on p. 66. This, in turn, is naturally isomorphic, by Theorem 1(viii), to $\mathrm{Ext}^{1}_{\mathrm{GrMod}\Lambda}(\Omega^{k+n-1+p}F(N)[p], J^{n-1}\Omega^{k}F(M)))$. Using Lemma 8 and dimension shift, the latter is naturally isomorphic to $\mathrm{Ext}^{n}_{\mathrm{GrMod}\Lambda}\left(\Omega^{k+p}F(N)[p], \Omega^{k}F(M)\right)$.

It remains to show that, for each nonnegative n, the constructed maps form a morphism of directed systems. We shall first do it in the case n = 0. Returning to the first chain of isomorphisms in the foregoing proof, notice that $\mathrm{Hom}_{GrModr}(J^{k}M, N[p])$ forms a directed over k system, for each value of p. The same is true for $\mathrm{Hom}_{GrModr}(J^{k}M, J^{k+p}N[p])$.

In each of these systems, a map f in the kth component of the system goes, using a slightly abused notation, to the map J f . It is clear that the first two isomorphisms in the above chain are morphisms of directed systems. Applying the functor F we make $\mathrm{Hom}_{GrMod\Lambda}(F(J^{k+p}N[p]), F(J^{k}M))$ into a directed over k system: the structure map sends F(f) to F(J f). Thus the third isomorphism in our chain becomes a morphism of directed systems. The next term $\mathrm{Hom}_{GrMod\Lambda}(\Omega^{k+p}F(N)[p], \Omega^{k}F(M))$ is also part of a directed system: a map g is sent to the map Ωg, the construction of Ω being explained in Remarks on p. 64. To show that the isomorphism

$$\mathrm{Hom}_{\mathrm{GrMod}\,\Lambda}(F(J^{k+p}N[p]), F(J^{k}M)) \cong \mathrm{Hom}_{\mathrm{GrMod}\,\Lambda}(\Omega^{k+p}F(N)[p], \Omega^{k}F(M))$$

is also a morphism of directed systems we need to show that the following diagram

$$\begin{array}{ccc}
\mathrm{Hom}_{\mathrm{GrMod}\,\Lambda}(F(J^{k+p}N[p]), F(J^{k}M)) & \longrightarrow & \mathrm{Hom}_{\mathrm{GrMod}\,\Lambda}(\Omega^{k+p}F(N)[p], \Omega^{k}F(M)) \\
\downarrow & & \downarrow \\
\mathrm{Hom}_{\mathrm{GrMod}\,\Lambda}(F(J^{k+1+p}N[p]), F(J^{k+1}M)) & \longrightarrow & \mathrm{Hom}_{\mathrm{GrMod}\,\Lambda}(\Omega^{k+1+p}F(N)[p], \Omega^{k+1}F(M))
\end{array}$$

commutes. The structure map sends $F(f) \in \mathrm{Hom}_{GrMod\Lambda}(F(J^{k+p}N[p]), F(J^{k}M))$ to $F(J f) \in \mathrm{Hom}_{GrMod\Lambda}(F(J^{k+1+p}N[p]), F(J^{k+1}M))$. The horizontal maps were constructed in the beginning of the proof. Applying the top map to F(f) we have F(f [k])[−k].

The structure map there is given by Ω, which was explicitly defined in remark (2) on p. 65. Applying it to F(f[k])[−k] and taking into account that Ω commutes with shifts, we obtain F(J f[k][1])[−1][−k]= F(J f[k + 1]) [−k − 1]. Applying the bottom map to F(J f) we obtain the same expression. This proves the compatibility for n = 0.

Assume now that n ≥ 1. In this case, the construction of the isomorphisms breaks down into the following steps:

$$\mathrm{Ext}^n_{\mathrm{Gr\,Mod}\,\Gamma}\left(J^kM, N[p]\right)$$
$$\Downarrow$$
$$\mathrm{Ext}^1_{\mathrm{Gr\,Mod}\,\Gamma}\left(\Omega^{n-1}J^kM, N[p]\right)$$
$$\Downarrow$$
$$\mathrm{Ext}^1_{\mathrm{Gr\,Mod}\,\Gamma}\left(\Omega^{n-1}J^kM, J^{k+n-1+p}N[p]\right)$$
$$\Downarrow$$
$$\mathrm{Ext}^1_{\mathrm{Gr\,Mod}\,\Gamma}\left(\Omega^{n-1}J^kM[k+n-1], J^{k+n-1+p}N[k+n-1+p]\right)$$
$$\Downarrow$$
$$\mathrm{Ext}^1_{\mathrm{Gr\,Mod}\,\Lambda}\left(F\left(J^{k+n-1+p}N[k+n-1+p]\right), F\left(\Omega^{n-1}J^kM[k+n-1]\right)\right)$$
$$\Downarrow$$
$$\mathrm{Ext}^1_{\mathrm{Gr\,Mod}\,\Lambda}\left(\Omega^{k+n-1+p}F(N)[k+n-1+p], J^{n-1}F\left(J^kM[k]\right)[n-1]\right)$$
$$\Downarrow$$
$$\mathrm{Ext}^1_{\mathrm{Gr\,Mod}\,\Lambda}\left(\Omega^{k+n-1+p}F(N)[k+n-1+p], J^{n-1}\Omega^kF(M)[k+n-1]\right)$$
$$\Downarrow$$
$$\mathrm{Ext}^1_{\mathrm{Gr\,Mod}\,\Lambda}\left(\Omega^{k+n-1+p}F(N)[p], J^{n-1}\Omega^kF(M)\right)$$
$$\Downarrow$$
$$\mathrm{Ext}^1_{\mathrm{Gr\,Mod}\,\Lambda}\left(\Omega^{k+n-1+p}F(N)[p], \Omega^kF(M)\right)$$
$$\Downarrow$$
$$\mathrm{Ext}^n_{\mathrm{Gr\,Mod}\,\Lambda}\left(\Omega^{k+p}F(N)[p], J^{n-1}\Omega^kF(M)\right)$$

Each entry above is part of a directed system. To show that the composition of the vertical maps is compatible with the directed systems on the top and on the bottom, it suffices to show that each vertical map is compatible with the directed systems on the source and on the target. While this verification, based on the standard properties of the Ext-functor, is tedious, it is straightforward and is left to the reader. This finishes the proof of the theorem.

Remark: Since a Koszul module is, by definition, of type FP_∞, we have, in the notation of the previous theorem, that $\mathrm{Ext}^n_\Gamma\left(J^kM,N\right) \cong \coprod_{p\in\mathbb{Z}} \mathrm{Ext}^n_{\mathrm{GrMod}\Gamma}\left(J^kM,N[p]\right)$ and, therefore,

$$\varinjlim_k \mathrm{Ext}^n_\Gamma(J^kM, N) \cong \coprod_{p\in\mathbb{Z}}\Big(\varinjlim_{k\geqslant\max\{-p,0\}} \mathrm{Ext}^n_{\mathrm{Gr\,Mod}\,\Lambda}\left(\Omega^{k+p}F(N)[p], \Omega^kF(M)\right)\Big)$$

This is an isomorphism of graded abelian groups, where the graded structure on each side is given by the index p.

Now we want specialize to the case n = 0 and work in the category of graded Λ-modules modulo projectives.More precisely, we want to compose, for each k, the established natural isomorphism

$$\tau : \mathrm{Hom}_{\mathrm{Gr\,Mod}\,\Gamma}(J^k M, N[p]) \to \mathrm{Hom}_{\mathrm{Gr\,Mod}\,\Lambda}(\Omega^{k+p} F(N)[p], \Omega^k F(M)) \quad (4.1)$$

with the canonical surjection

$$\pi : \mathrm{Hom}_{\mathrm{Gr\,Mod}\,\Lambda}(\Omega^{k+p} F(N)[p], \Omega^k F(M)) \to \underline{\mathrm{Hom}}_{\mathrm{Gr\,Mod}\,\Lambda}(\Omega^{k+p} F(N)[p], \Omega^k F(M)) \quad (4.2)$$

The obtained composition

$$\pi\tau : \mathrm{Hom}_{\mathrm{Gr\,Mod}\,\Gamma}(J^k M, N[p]) \to \underline{\mathrm{Hom}}_{\mathrm{Gr\,Mod}\,\Lambda}(\Omega^{k+p} F(N)[p], \Omega^k F(M)) \quad (4.3)$$

is obviously natural.

It is also clear that the canonical maps (4.2) form a morphismof directed over k systems. As a result, the composition (4.3) becomes a morphism of directed systems. Our next goal is to show that, in the limit, this morphism induces an isomorphism. First, we recall an auxiliary result [9, Proposition 4.2].

Proposition 1: Let Γ be a graded quiver algebra, M and N graded Γ–modules, and let t (N) denote the sum of all submodules of N of finite length. If M and all graded simple Γ–modules are of type FP_∞, then the map

$$\varinjlim_k \mathrm{Ext}^i_\Gamma(M_{\geqslant k}, N) \to \varinjlim_k \mathrm{Ext}^i_\Gamma(M_{\geqslant k}, N/t(N))$$

induced by the canonical surjection N →N/t (N), is an isomorphism for each I ≥ 0. We can now state and prove the first main result of the paper.

Theorem 4: Let Γ be a Koszul algebra with radical J and Yoneda algebra Λ and let $F : \mathrm{mod}\Gamma \to \mathrm{mod}\Lambda$ be the Koszul duality:

$F(-)\coprod_{i\geq 0} = \mathrm{Ext}^i_\Gamma(-,\Gamma/J)$. Then, for any pair of Koszul Γ –modules M and N and any integer p, the just constructed composition $\pi\tau$ gives rise to a natural isomorphism

$$\varinjlim_{k\geqslant 0} \mathrm{Hom}_{\mathrm{GrMod}\,\Gamma}(J^k M, N[p]) \xrightarrow{\cong} \varinjlim_{k\geqslant \max\{-p,0\}} \underline{\mathrm{Hom}}_{\mathrm{GrMod}\,\Lambda}(\Omega^{k+p}F(N)[p], \Omega^k F(M)). \tag{4.4}$$

Proof. The surjectivity of the map is obvious. To show the injectivity, let f be a representative of an element in the kernel of this map. Then, for a large enough k, (a descendant of) the map $\pi\tau(f) \in \mathrm{Hom}_{\mathrm{GrMod}\Lambda}(\Omega^{k+p}F(N)[p],\Omega_k F(M))$ will vanish. Thismeans that $\tau(f) \in \mathrm{Hom}_{\mathrm{GrMod}\Lambda}(\Omega^{k+p}F(N)[p],\Omega_k F(M))$ factors through a projective. Since Koszul duality interchanges projectives and semisimples, $f \in \mathrm{Hom}_{\mathrm{GrMod}\Gamma}(J^kM,N[p])$ factors through a semisimple module. Thus the image of f is a submodule of t (N). Since Γ and M are Koszul, the assumptions of the previous proposition hold. Moreover, as M is generated in the highest degree zero, $M_{\geq k} = J^kM$. Thus the canonical surjection $N \to N/t(N)$ induces an isomorphism

$$\varinjlim_{k} \mathrm{Hom}_{\mathrm{GrMod}\,\Gamma}(J^k M, N[p]) \xrightarrow{\cong} \varinjlim_{k} \mathrm{Hom}_{\mathrm{GrMod}\,\Gamma}(J^k M, N/t(N)[p])$$

Under this isomorphism, the class of f goes to zero. Therefore the class of f is zero.

COHOMOLOGY OF TAILS AND VOGEL COHOMOLOGY

In this section we shall interpret Theorem 4 in terms of quotient categories and of Vogel cohomology. Let Γ be a graded quiver algebra with graded radical J, and GrModΓ the category of graded Γ -modules. The symbol FlΓ (respectively, flΓ) will denote the full subcategory of GrModΓ (respectively, GrmodΓ) consisting of modules whose elements are annihilated by powers of J . It is immediate that FlΓ is a Serre subcategory, and we shall refer to modules in FlΓ as torsion modules (more precisely, such modules should be called J -torsion modules). Any finitely generated torsion module is obviously of finite length. Since any module is the direct limit of its finitely generated submodules, torsion modules can also be characterized as direct limits of modules of finite length.

Being a Serre subcategory, FlΓ gives rise to a quotient category GrModΓ/FlΓ in the sense of Serre [4]. It is sometimes called the category of tails of Γ-modules, denoted TailsΓ, and is defined as follows. The objects of TailsΓ are the same as of GrModΓ and

$$\mathrm{Hom}_{\mathrm{Tails}\,\Gamma}(M, N) := \varinjlim_{k} \mathrm{Hom}_{\mathrm{Gr\,Mod}\,\Gamma}(M', N/N')$$

Here the limit is taken over all pairs (M′,N′), where M′ is a sub module of M and N′ is a submodule of N such that M/M′ and N′ are torsion, and the pairs are ordered as follows: (M′,N′) ≤ (M″,N″) if and only if M″ ⊆M′ and N′ ⊆ N″. The category Tails Γ is abelian with enough injectives, and the canonical functor π :GrModΓ →TailsΓ is exact. A Γ–module M, viewed as an object of Tails Γ will be denoted M and sometimes referred to as the tail of M. An example of fundamental importance is provided by a theorem of Serre (see [12]) which asserts, in the present terminology, that the category of coherent sheaves on a projective space is equivalent to the category of tails of finitely generated graded modules over the corresponding polynomial algebra. In this case the Serre subcategory consists of graded modules of finite length.

Under mild assumptions, the description of morphisms in Tails Γ can be simplified. As before, let t (N) denote the sum of all sub-modules of N of finite length. We then have [9, Proposition 4.2]

Proposition 2: Let Γ be a graded quiver algebra and M and N graded Γ-modules. If M is finitely generated, then there is a natural isomorphism

$$\mathrm{Hom}_{\mathrm{Tails}\,\Gamma}(\widetilde{M}, \widetilde{N}) \cong \varinjlim_{k} \mathrm{Hom}_{\mathrm{Gr\,Mod}\,\Gamma}(M_{\geqslant k}, N/t(N))$$

If M and all graded simples are of type FP_∞, then, according to Proposition 1 on p. 70, the canonical surjection N →N/t (N) induces an isomorphism

$$\varinjlim_{k} \mathrm{Hom}_{\mathrm{Gr\,Mod}\,\Gamma}(M_{\geqslant k}, N) \xrightarrow{\cong} \varinjlim_{k} \mathrm{Hom}_{\mathrm{Gr\,Mod}\,\Gamma}\left(M_{\geqslant k}, N/t(N)\right)$$

Combining these results with the fact that, over a Koszul algebra, all Koszul modules are, by definition, of type FP_∞, we now have the following interpretation of the left-hand side of the isomorphism in the statement of Theorem 4 on p. 70.

Proposition 3: Let Γ be a Koszul algebra with graded radical J . Then, for any Koszul Γ -modules M and any Γ -module N, there is a natural isomorphism

$$\varinjlim \mathrm{Hom}_{\mathrm{Gr\,Mod}\,\Gamma}\left(J^k M, N[p]\right) \cong \mathrm{Hom}_{\mathrm{Tails}\,\Gamma}\left(\widetilde{M}, \widetilde{N}[p]\right)$$

In other words, the left-hand side of the isomorphism in Theorem 4 is naturally isomorphic to the degree zero "tail cohomology" of $\tilde{M}$ with coefficients in $\tilde{N}$ [p].

To interpret the right-hand side, we recall the definition of (a graded version of) Vogel cohomology (also known as Tate–Vogel cohomology), due to the eponymous author, which extends Tate cohomology to arbitrary associative rings with identity (for more details, see [11]). Let Λ be an arbitrary graded ring, and M and N arbitrary graded Λ-modules. Let (P_M, d_M), respectively, (P_N, d_N), be a graded projective resolution of the Λ-module M, respectively, N. The symbol (P_M, P_N) will indicate the graded Λ_0-module whose degree n component consists of bidegree (n, 0) bihomogeneous maps from P_M to P_N. The degree n here refers to the dimension shift in the projective resolutions, and not to the graded structure of each projective. In the latter grading, the maps are still required to be of degree zero. The symbol $(P_M, P_N)b$ will indicate the graded Λ_0-submodule consisting of boundedmaps, i.e., the bihomogeneous maps with only finitely many nonzero components (with respect to the first degree). Let $\underline{(P_M, P_N)}$ be the corresponding quotient module. As a result, we have a short exact sequence

$$0 \to (\mathbf{P}_M, \mathbf{P}_N)_b \to (\mathbf{P}_M, \mathbf{P}_N) \xrightarrow{\pi} \underline{(\mathbf{P}_M, \mathbf{P}_N)} \to 0$$

of graded (not bigraded!) Λ_0-modules. The standard formula $D(f) := d_N \circ f - (-1)^{\deg f} f \circ d_M$ defines a differential on the middle module, which obviously restricts to the submodule of bounded maps. Thus we have a short exact sequence of complexes and graded Vogel cohomology $\overline{\operatorname{Ext}}^i_{\mathrm{GrMod}\Lambda}(M,N)$ is defined as the $-i$ th homology $H_{-i}(P_M, P_N)$ of the quotient complex. In that complex, the degree zero chains are bihomogeneous maps of degree (0, 0), the degree zero cycles are the bihomogeneous maps of degree (0, 0) which commute with the differentials d_M and d_N in all high enough homological degrees, and the degree zero boundaries are the degree zero cycles which are null-homotopic in all high enough degrees.5

Our next goal is to show that Vogel cohomology can be obtained by "stabilizing" the syzygy endofunctor. More precisely, we want to establish a natural isomorphism

$$\varinjlim_{k \geqslant \max\{-p,0\}} \underline{\operatorname{Hom}}_{\mathrm{Gr\,Mod}\,\Lambda}(\Omega^{k+p}M, \Omega^k N) \xrightarrow{\cong} \overline{\operatorname{Ext}}^p_{\mathrm{Gr\,Mod}\,\Lambda}(M, N) \tag{5.1}$$

This result was first obtained, over Gorenstein rings, by Buchweitz (see [2]). To construct the desired isomorphism, we choose an arbitrary morphism $\tilde{\alpha}_k \in \underline{\operatorname{Hom}}_{\mathrm{GrMod}\Lambda}(\Omega^{k+p}M, \Omega^k N)$ and an arbitrary representative $(\alpha_k : \Omega^{k+p}M \to \Omega^k N) \in \tilde{\alpha}_k$. Lifting α_k to the corresponding projective resolutions, we have a bidegree $(-p, 0)$ cycle in (P_M, P_N). Any morphism that factors through a projective is sent to zero by this map. Thus, defining the image of $\tilde{\alpha}_k$ as the image of α_k, we have a well-defined map into the bidegree $(-p, 0)$ cycles. Finally, composing this map with the surjection to the homology, we have, for each value of $k \geq 0$, a well-defined map $\underline{\operatorname{Hom}}_{\mathrm{GrMod}\Lambda}(\Omega^{k+p}M, \Omega^k N) \to \overline{\operatorname{Ext}}^p_{\mathrm{GrMod}\Lambda}(M,N)$.

It is clear that these maps are compatible with Ω, and, therefore, we obtain a map as in (5.1). It is routine to check that this map is a natural isomorphism. The details are left to the reader.

We can now reformulate Theorem 4.

Theorem 5: Let Γ be a Koszul algebra with Yoneda algebra Λ and let F: modΓ →modΛ be the Koszul duality: $F(-)\coprod_{i\geq 0} = \mathrm{Ext}^i_\Gamma(-,\Gamma/J)$. Then, for any pair of Koszul Γ-modules M and N and any integer p, there is a natural isomorphism

$$\mathrm{Hom}_{\mathrm{Tails}\,\Gamma}(\widetilde{M},\widetilde{N}[p]) \xrightarrow{\cong} \underline{\mathrm{Ext}}^p_{\mathrm{Gr\,Mod}\,\Lambda}(F(N)[p],F(M))$$

between the degree zero tail cohomology over Γ and Vogel cohomology over Λ.

Our next goal is to extend the last result to higher-order tail cohomology. As a first step, we identify the Ω-stabilization of the higher Ext-functors.

Lemma 9: Let Λ be any graded ring, and M and N graded Λ-modules. Then, for any

integer n ≥ 1 and any integer p, there is a natural isomorphism

$$f:\quad \varinjlim_{k\geqslant \max\{-p,0\}} \mathrm{Ext}^n_{\mathrm{Gr\,Mod}\,\Lambda}(\Omega^{k+p}M,\Omega^k N) \xrightarrow{\cong} \underline{\mathrm{Ext}}^{n+p}_{\mathrm{Gr\,Mod}\,\Lambda}(M,N)$$

Proof: Let P→M and Q→N be projective resolutions. To construct f we choose a representative $\alpha: P_{\geq n+p+k} \to Q_{\geq k}$ of $\tilde{\alpha} \in \mathrm{Ext}^n_{\mathrm{GrMod}\Lambda}(\Omega^{k+p}M,\Omega^k N)$. Then α is a chain map of bidegree (−n−p, 0). Extending it by zero, we have a bidegree (−n−p, 0) map of graded modules α' :**P**→**Q**. We nowdefine $f_n(\tilde{\alpha})$ as the class of α in $\mathrm{Ext}^{n+p}_{\mathrm{GrMod}\Lambda}(M,N)$.

Any other representative of $\tilde{\alpha}$ is homotopic to α and, therefore, the map f_n is well-defined.

Any two choices for $\Omega\alpha$ are homotopic and, therefore, $(\Omega\alpha)' - \alpha'$ is homotopic to a bounded map. This shows that $f_{n+1}(\Omega\tilde{\alpha}) = f_n(\tilde{\alpha})$ and, passing to the limit, we have the desired map f . To show that f is an isomorphism, we shall construct an inverse g of f. Let $\beta : \mathbf{P} \to \mathbf{Q}$ be a representative of $\tilde{\beta} \in \mathrm{Ext}^{n+p}_{\mathrm{GrMod}\Lambda}(M,N)$. This means that β is a bidegree (−n − p, 0) map of graded modules which commutes with the differentials in all high enough degrees. In other words, $\beta_{\geq n+p+k} : \mathbf{P}_{\geq n+p+k} \to \mathbf{Q}_{\geq k}$ is a chain map for each large enough k. Therefore $\beta_{\geq n+p+k}$ gives rise to an element of $\mathrm{Ext}^{n}_{\mathrm{GrMod}\Lambda}(\Omega^{k+p}M, \Omega^{k}N)$ for each large enough k. All these elements have the same image γ in the direct limit and we set $g(\beta) := \gamma$. It is clear that g is well-defined and that fg and gf are the identity maps.

As a second step, before we pass to the higher-order tail cohomology, we recall another auxiliary result [9, Proposition 4.5].

Proposition 4: Let Γ be a graded quiver algebra such that all graded simples are of type FP_∞, M a graded Γ-module of type FP_∞, and N any graded Γ-module. Then $\mathrm{Ext}^{n}_{\mathrm{Tails}\,\Gamma}(\tilde{M},\tilde{N})$ is naturally isomorphic to $\varinjlim \mathrm{Ext}^{n}_{\mathrm{GrMod}\Gamma}(M_{\geq k}, N)$ for any integer $n \geq 0$. Now we can state the second main result of the paper.

Theorem 6: Let Γ be a Koszul algebra with Yoneda algebra Λ and let $F : \mathrm{mod}\Gamma \to \mathrm{mod}\Lambda$ be the Koszul duality: $F(-) = \coprod_{i\geq 0} \mathrm{Ext}^{i}_{\Gamma}(-, \Gamma/J)$. Then, for any pair of Koszul Γ–modules M and N, any integer p, and any integer $n \geq 0$ there is a natural isomorphism

$$\mathrm{Ext}^{n}_{\mathrm{Tails}\,\Gamma}(\widetilde{M}, \widetilde{N}[p]) \xrightarrow{\cong} \underline{\mathrm{Ext}}^{n+p}_{\mathrm{Gr\,Mod}\,\Lambda}(F(N)[p], F(M))$$

between the degree n tail cohomology over Γ and the degree n + p Vogel cohomology over Λ.

Proof: The case $n=0$ has already been proved above. Suppose that $n \geq 1$. By Proposition 4 on this page, $\mathrm{Ext}^{n}_{\mathrm{Tails}\Gamma}(\tilde{M}, \tilde{N}[p]) \cong \varinjlim \mathrm{Ext}^{n}_{\mathrm{GrMod}\Gamma}(M_{\geq k}, N[p])$. Since M is generated in the highest degree zero, $M_{\geq k} = J^{k}M$. Thus the left-hand side is isomorphic to $\varinjlim \mathrm{Ext}^{n}_{\mathrm{GrMod}\Gamma}(J^{k}M, N[p])$. The desired result now follows from Theorem 3 on p. 67 and Lemma 9 on p. 73. _

Our next goal is to impose certain restrictions on the algebras in question and investigate the just stated theorem under the new assumptions. Let Γ be a Koszul algebra and $K\Gamma$ the full subcategory of $\mathrm{Grmod}\Gamma$ consisting of Koszul modules. Furthermore, let $[K\Gamma]$ denote the full subcategory of $\mathrm{Grmod}\Gamma/\mathrm{fl}\Gamma =: \mathrm{tails}\Gamma$ whose objects are arbitrary shifts of Koszul Γ-modules. We now have a commutative diagram of functors

$$\begin{array}{ccc} K_\Gamma & \hookrightarrow & \mathrm{Gr\,mod}\,\Gamma \\ \downarrow & & \downarrow \\ [\tilde{K}_\Gamma] & \overset{I}{\hookrightarrow} & \mathrm{Gr\,mod}\,\Gamma/\mathrm{fl}\,\Gamma = \mathrm{tails}\,\Gamma, \end{array}$$

where the horizontal functors are inclusions.

Proposition 5: Suppose that Γ is a Noetherian Koszul algebra of finite global dimension.

Then the functor

$I : [\tilde{K}_\Gamma] \hookrightarrow \mathrm{Gr\,mod}\,\Gamma/\mathrm{Fl}\,\Gamma = \mathrm{tails}\,\Gamma$

is an equivalence of categories.

Proof: By construction, I is full and faithful. It remains to show that it is dense. But this follows immediately from Theorem 2 on p. 66.

Now let Λ be the Yoneda algebra of Γ. We shall define the (graded) Vogel category $v(\Lambda)$ of Λ. Its objects are finitely generated graded Λ-modules. If M and N are such modules, we define $\mathrm{Hom}_{v(\Lambda)}(M,N)$ as

$$\underline{\mathrm{Ext}}^0_{\mathrm{Gr\,Mod}\,\Lambda}(M,N) = \varinjlim_{k \geqslant 0} \underline{\mathrm{Hom}}_{\mathrm{Gr\,Mod}\,\Lambda}\left(\Omega^k M, \Omega^k N\right)$$

Proposition 6: Suppose that Γ is a Noetherian Koszul algebra of finite global dimension with Yoneda algebra Λ. Let $\left[\tilde{K}_\Gamma\right]_n, n \in \mathbb{Z}$, denote the full subcategory of $\left[\tilde{K}_\Gamma\right]$ whose objects are generated in degree n. Then, for any $n \in \mathbb{Z}$, the functor $F: K_\Gamma \to K_\Lambda$ induces a faithful and full contravariant functor $\left[\tilde{F}\right]_n : \left[\tilde{K}_\Gamma\right]_n \to v(\Lambda)$. The image of this functor consists of all Koszul Λ-modules.

Proof: This is just a restatement of the previous proposition combined with Theorem 5 of this section.

TATE COHOMOLOGY OVER SELFINJECTIVE RINGS

In this section we want to give a streamlined construction of Tate cohomology, which is a particular case of Vogel cohomology. Throughout the rest of this section, we assume that Λ is a selfinjective graded Noetherian ring. In this case, the construction of graded Tate cohomology can be made particularly simple. It works equally well in the non-graded case. For our purposes we want to consider only finitely generated graded (left) Λ-modules. Let M be such a module. From the exact sequence (see [1, Proposition IV.3.2])

$$0 \to \mathrm{Ext}^1_\Lambda(\mathrm{Tr}\, M, \Lambda) \to M \xrightarrow{e_M} M^{**} \to \mathrm{Ext}^2_\Lambda(\mathrm{Tr}\, M, \Lambda) \to 0,$$

where $(-)^* = \mathrm{Hom}_\Lambda(-,\Lambda)$ and e_M is the canonical evaluation map, it follows immediately that M is reflexive. It is easy to check that the map e_M is of degree zero. Let

$$\mathbf{P} \quad \cdots \to P_1 \to P_0 \to M \to 0$$

be a graded projective resolution of M and

$$\mathbf{Q} \quad \cdots \to Q_1 \to Q_0 \to M^* \to 0$$

a graded projective resolution of M^*. The degree zero isomorphism e_M gives rise to a bidegree (0, 0) quism $\mathbf{P} \to \mathbf{Q}^*$. The mapping cone of it, shifted by −1, is called a complete resolution of M. This is an exact complex of graded projectives and bidegree (−1, 0) differentials. If M is not projective, it is not contractible. Using this complete resolution of M in place of a graded projective resolution of M, we obtain graded Tate cohomology groups $\underline{\mathrm{Ext}}^{i}_{\mathrm{GrMod}\Lambda}(M,-)$, which are naturally isomorphic (see, for example, [11]), as the notation suggests, to Vogel cohomology. It is immediate that, under our assumptions, $\underline{\mathrm{Ext}}^{i}_{\mathrm{GrMod}\Lambda}(M,-) \cong \mathrm{Ext}^{i}_{\mathrm{GrMod}\Lambda}(M,-)$ for all $i \geq 1$. Furthermore, since Λ is self-injective, the syzygy operation Ω is an endofunctor on and, in fact, an auto-equivalence of the category of Λ-modules modulo projectives (see [7]). Now it is not difficult to check that $\underline{\mathrm{Ext}}^{i}_{\mathrm{GrMod}\Lambda}(M,N)$ is naturally isomorphic to $\underline{\mathrm{Ext}}^{i}_{\mathrm{GrMod}\Lambda}(\Omega M,\Omega N)$ for all integers i. Furthermore, the new cohomology admits infinite dimension shifts in both directions for each of the two variables. In particular, $\underline{\mathrm{Ext}}^{i}_{\mathrm{GrMod}\Lambda}(M,N)$ is naturally isomorphic to $\underline{\mathrm{Ext}}^{i+1}_{\mathrm{GrMod}\Lambda}(M,\Omega N)$. Next we want to identify Tate cohomology groups in nonpositive degrees.

Lemma 10: Under the above assumptions on Λ, if M and N are graded Λ-modules and Nis finitely generated, then $\underline{\mathrm{Hom}}_{\mathrm{GrMod}\Lambda}(M,N)$ is naturally isomorphic to $\underline{\mathrm{Ext}}^{1}_{\mathrm{GrMod}\Lambda}(M,\Omega N)$, where $\underline{\mathrm{Hom}}_{\mathrm{GrMod}\Lambda}(M,N)$ denotes the morphisms from M to N in the category of graded Λ-modules modulo projectives.

Proof: Applying the functor $\mathrm{Hom}_{\mathrm{GrMod}\Lambda}(M,-)$ to the exact sequence $0 \to \Omega N \to P \to N \to 0$, where P is a finite projective precover, we have the desired result.

Using dimension shift in the covariant argument, we have, as a consequence of the lemma, a natural isomorphism $\underline{\mathrm{Ext}}^{0}_{\mathrm{GrMod}\Lambda}(M,N) \cong \underline{\mathrm{Hom}}_{\mathrm{GrMod}\Lambda}(M,N)$. The latter is known, if M is finitely presented (see [10, Proposition 1.6.3]) to be naturally isomorphic as a graded Λ_0-module to $\mathrm{Tor}_1^{\mathrm{GrMod}\Lambda}(\mathrm{Tr}M,N)$, where TrM denotes a graded transpose of M.6 (For details on graded transpose, see Section 1.4 of [10].)

This identifies Tate cohomology groups in degree zero. The remaining groups can be identified by using dimension shift in the covariant argument. More precisely, if i ≥ 0, then $\underline{\mathrm{Ext}}^{-i}_{\mathrm{GrMod}\Lambda}(M,N) \cong \underline{\mathrm{Ext}}^{0}_{\mathrm{GrMod}\Lambda}(M,\Omega^i N)$. By the above, this is naturally isomorphic to

$$\mathrm{Tor}_1^{\mathrm{Gr\,Mod}\,\Lambda}(\mathrm{Tr}\,M, \Omega^i N) \cong \mathrm{Tor}_1^{\mathrm{Gr\,Mod}\,\Lambda}(\Omega^i\,\mathrm{Tr}\,M, N)$$

Putting together the proved results, we have

Theorem 7: Let Λ be a graded Noetherian selfinjective ring and M and N finitely generated graded Λ-modules. Then there are natural isomorphisms:

- $\underline{\mathrm{Ext}}^{i}_{\mathrm{Gr\,Mod}\,\Lambda}(\Omega^k M, N) \cong \underline{\mathrm{Ext}}^{i+k}_{\mathrm{Gr\,Mod}\,\Lambda}(M, N)$ *for all* $i, k \in \mathbb{Z}$.
- $\underline{\mathrm{Ext}}^{i}_{\mathrm{Gr\,Mod}\,\Lambda}(M, \Omega^k N) \cong \underline{\mathrm{Ext}}^{i-k}_{\mathrm{Gr\,Mod}\,\Lambda}(M, N)$ *for all* $i, k \in \mathbb{Z}$.
- $\underline{\mathrm{Ext}}^{i}_{\mathrm{Gr\,Mod}\,\Lambda}(\Omega^k M, \Omega^k N) \cong \underline{\mathrm{Ext}}^{i}_{\mathrm{Gr\,Mod}\,\Lambda}(M, N)$ *for all* $i, k \in \mathbb{Z}$.
- $\underline{\mathrm{Ext}}^{i}_{\mathrm{Gr\,Mod}\,\Lambda}(M, N) \cong \mathrm{Ext}^{i}_{\mathrm{Gr\,Mod}\,\Lambda}(M, N)$ *for all* $i \geqslant 1$.
- $\underline{\mathrm{Ext}}^{-i}_{\mathrm{Gr\,Mod}\,\Lambda}(M, N) \cong \mathrm{Tor}_1^{\mathrm{Gr\,Mod}\,\Lambda}(\Omega^i\,\mathrm{Tr}\,M, N) \cong \underline{\mathrm{Hom}}_{\mathrm{Gr\,Mod}\,\Lambda}(\mathrm{Tr}\,\Omega^i\,\mathrm{Tr}\,M, N)$ *for all* $i \geqslant 0$.
- $\underline{\mathrm{Ext}}^{0}_{\mathrm{Gr\,Mod}\,\Lambda}(M, N) \cong \underline{\mathrm{Hom}}_{\mathrm{Gr\,Mod}\,\Lambda}(M, N) \cong \mathrm{Tor}_1^{\mathrm{Gr\,Mod}\,\Lambda}(\mathrm{Tr}\,M, N)$.

Remark: Let Λ be a selfinjective gradedNoetherian ring andM a graded finitely generated Λ-module. If M has a projective cover, then the kernel of this cover has no projective summands.

SELFINJECTIVE KOSZUL ALGEBRAS AND THE NAKAYAMA AUTOMORPHISM

In this section we shall study the relationship between the automorphism σ , which appears in the Serre duality formula for generalized Artin–Schelter regular algebras (see [9] and [10] for details), and the Nakayama automorphism of its Yoneda algebra.

We begin by recalling some facts about selfinjective graded quiver algebras. The proofs of the next four lemmas can be obtained by an easy adaptation of the standard proofs for nongraded algebras (see, for example, [15]). For the first lemma, notice that any basic selfinjective graded algebra Λ of Loewy length, say, n is isomorphic, as a left Λ-module, to $\mathrm{Homgrk}({}_k\Lambda_\Lambda, k)[-n]$. In otherwords, Λ is necessarily graded Frobenius. In particular, this applies to graded quiver algebras.

Lemma 11: Let Λ be a basic selfinjective graded algebra of Loewy length n and $\vartheta : {}_\Lambda\Lambda \to D(\Lambda_\Lambda)[-n]$ an isomorphism of graded modules determined by the duality $D(-) := \mathrm{Homgr}_k(-,k)$. Then the bilinear form $f = \{f_j\} : \Lambda \times \Lambda[n] \to k$ with $f_j : \Lambda_j \times \Lambda_{n-j} \to k$, defined by $f(x, y) := \vartheta(y)x$ is nondegenerate and $f(xz,y) = f(x,zy)$ for all $x, y, z \in \Lambda$.

Lemma 12: Under the assumptions of Lemma 11, there exists a graded k-algebra automorphism $\nu: \Lambda \to \Lambda$, called the Nakayama automorphism, such that $f(x, y) = f(\nu(y), x)$ for all $x, y \in \Lambda$.

Given any k-algebra automorphism $\tau: \Lambda \to \Lambda$ and any left (respectively, right) Λ-module M we can modify ("twist") the module structure of M by defining $\lambda \cdot m$ (respectively, $m \cdot \lambda$) as $\tau(\lambda)m$ (respectively, $m\tau(\lambda)$) for any $\lambda \in \Lambda$ and $m \in M$. The resulting module will

be denoted ${}^\tau M$ (respectively, M^τ). It is immediate that $(-)_\tau$ and ${}_\tau(-)$ are autoequivalences of the category of gradedΛ-modules. In particular, both functors are exact. Notice also that both functors preserve the Yoneda multiplication. Modifying the right Λ-module structure of $D(\Lambda_\Lambda)$ $[-n]$, we have

Lemma 13: Under the assumptions of Lemma 11, the map $\vartheta: {}_\Lambda\Lambda \to D(\Lambda_\Lambda)$ $[-n]^\nu$ is a graded bimodule isomorphism.

Recall that the composition of dualities

$$\mathcal{N} : \mathrm{LFGr\,Mod}\,\Lambda \xrightarrow{\mathrm{Homgr}_\Lambda(?,{}_\Lambda\Lambda)} \mathrm{LFGr\,Mod}\,\Lambda^{\mathrm{op}} \xrightarrow{D(?)[-n]} \mathrm{LFGr\,Mod}\,\Lambda$$

is called the (graded) Nakayama equivalence. The next lemma gives relations between the Nakayama equivalence and the Nakayama automorphism. In colloquial terms, the Nakayama equivalence comes from the Nakayama automorphism.

Lemma 14: Let $\Lambda = kQ/I$ be a selfinjective graded quiver algebra of Loewy length n, ν a Nakayama automorphism of Λ, and M a locally finite graded Λ-module. Then there are natural isomorphisms

$$D(M^*)[-n] \cong {}^{\nu}M \quad and \quad (D(M))^*[n] \cong {}^{\nu^{-1}}M.$$

where $(-)^*$ denotes $\mathrm{Homgr}_\Lambda(-,\Lambda)$ and, respectively, $\mathrm{Homgr}_{\Lambda^{op}}(-,\Lambda^{op})$.

Proof: By the adjointness of the Hom-functor and the tensor product we have

$$\mathrm{Homgr}_{\Bbbk}(D(\Lambda)[-n]\otimes_\Lambda M, \Bbbk) \cong \mathrm{Homgr}_\Lambda(M, D^2(\Lambda)[n]) \cong \mathrm{Homgr}_\Lambda(M,\Lambda)[n] = M^*[n].$$

Applying D(–) again, we have $D(\Lambda)[-n]\otimes_\Lambda M \cong D(M^*)[-n]$. On the other hand, Lemma 13 shows that $D(\Lambda)[-n]$ is isomorphic to $\Lambda^{\nu^{-1}}$. Consequently,

$$D(M^*)[-n] \cong \Lambda^{\nu^{-1}} \otimes_\Lambda M \cong \Lambda \otimes_\Lambda {}^{\nu}M \cong {}^{\nu}M$$

This establishes the first assertion. The second assertion is a consequence of the following isomorphisms:

$$\begin{aligned}(D(M))^*[n] &\cong \mathrm{Homgr}_{\Lambda^{op}}(D(M)[-n],\Lambda) \cong \mathrm{Homgr}_\Lambda(D(\Lambda), M[n])\\ &\cong \mathrm{Homgr}_\Lambda(D(\Lambda)[-n], M).\end{aligned}$$

As $D(\Lambda)[-n] \cong \Lambda^{\nu^{-1}}$, the latter is isomorphic to $\mathrm{Homgr}_\Lambda(\Lambda^{\nu^{-1}}, M)$, which is

$${}^{\nu^{-1}}\mathrm{Homgr}_\Lambda(\Lambda, M) \cong {}^{\nu^{-1}}M.$$

The lemma is proved.

Our next goal is to investigate how a twist by a graded algebra automorphism affects the corresponding Yoneda algebra. The answer is given by the following

Proposition 7: Let Λ be a nonnegatively graded k-algebra with Yoneda algebra $\Gamma = \coprod_{k\geq 0} \mathrm{Ext}^k_\Lambda(\Lambda_0,\Lambda_0)$ and $\sigma: \Lambda\to\Lambda$ a graded algebra automorphism. Then σ induces a graded algebra automorphism $\rho: \Gamma \to\Gamma$ such that, for any Λ-moduleM, $F({}^\sigma M)$ is naturally isomorphic to $\rho^{-1}F(M)$, where $F(-) = \coprod_{k\geq 0} \mathrm{Ext}^k_\Lambda(-,\Lambda_0)$ is the functor defined in Section 2, Theorem 1(viii).

Proof: Suppose $x \in \mathrm{Ext}^k_\Lambda(\Lambda_0,\Lambda_0), k \geq 1$, is represented by a k-fold extension

$$x : 0 \to \Lambda_0 \xrightarrow{i} E_1 \to \cdots \to E_k \xrightarrow{p} \Lambda_0 \to 0$$

Since ${}^\sigma(-)$ is an exact functor, twisting each term of x by σ we obtain a new k-fold extension ${}^\sigma x$

$${}^\sigma x : 0 \to {}^\sigma\Lambda_0 \xrightarrow{{}^\sigma i} {}^\sigma E_1 \to \cdots \to {}^\sigma E_k \xrightarrow{{}^\sigma p} {}^\sigma\Lambda_0 \to 0$$

The map $\psi := {}_\sigma|\Lambda_0 : \Lambda_0 \to {}^\sigma\Lambda_0$ is obviously an isomorphismof Λ-modules. Let $\rho_k(x) := \psi^{-1}\acute{e}(\sigma x)\acute{e}\psi \in \mathrm{Ext}^k_\Lambda(\Lambda_0,\Lambda_0)$ be the k-fold extension obtained by conjugating ${}^\sigma x$ by ψ using the Yoneda multiplication. Then the map $\rho := (\rho_k)_{k\geq 0} : \Gamma \to\Gamma$ is a graded kalgebra automorphism with $\rho^{-1} = (\rho^{-1k})_{k\geq 0}$, where

$$\rho_k^{-1}(x) = {}^{\sigma^{-1}}\left(\psi \circ x \circ \psi^{-1}\right).$$

Next we construct, for each nonnegative value of k, a k-vector space isomorphism $\phi_k : \mathrm{Ext}^k_\Lambda({}^\sigma M,\Lambda_0) \to \mathrm{Ext}^K_\Lambda(M,\Lambda_0)$. For an extension $y \in \mathrm{Ext}^k_\Lambda({}^\sigma M,\Lambda_0)$, define $\varphi_k(y)$ as

$$\left({}^{\sigma^{-1}}\psi\right) \circ \left({}^{\sigma^{-1}}y\right) = {}^{\sigma^{-1}}(\psi \circ y).$$

This map is an isomorphism with the inverse $z \mapsto \psi^{-1}\acute{e}({}^\sigma z)$.We claim that $\varphi := (\varphi_k)_{k\geq 0}$ is a homomorphism of the corresponding Γ-modules.

Choose $x \in \mathrm{Ext}^j_\Lambda(\Lambda_0, \Lambda_0)$ and $y \in \mathrm{Ext}^k_\Lambda({}^\sigma M, \Lambda_0)$. We need to show that $\varphi_{k+j}(x\acute{e}y) = \rho_j^{-1}(x)\acute{e}\varphi_k(y)$. But this immediately follows from the definitions of φ and ρ^{-1}. The naturality of the just constructed isomorphism is obvious.

Next we want to recall [9, p. 244] the definition of generalized Artin–Schelter regular

algebras.7

Definition 2: A graded quiver algebra Λ is said to be generalized Artin–Schelter regular of dimension n if:

(1) All graded simples are of type FP_∞ and of projective dimension n.

(2) If S is a graded simple, then $\mathrm{Ext}^k_\Lambda(S, \Lambda) = 0$ for $0 \le k \le n-1$.

(3) The functor $\mathrm{Ext}^n_\Lambda(-, \Lambda)$ induces a bijection between the isomorphism classes of graded simple Λ- and graded simple Λ^{op}-modules.

The importance of this class of algebras is explained by the following result [9, Theorem 3.1].

Theorem 8: Let Λ be an indecomposable finite-dimensional Koszul quiver algebra. Then Λ is selfinjective if and only if its Yoneda algebra is generalized Artin–Schelter regular.

Assume now that Λ is an indecomposable selfinjective Koszul quiver algebra of Loewy length n _ 2 with Yoneda algebra Γ. By the previous theorem, Γ is generalized Artin–Schelter regular. Moreover, a noncommutative version of the Serre duality holds for Γ [9,Theorem 4.9]. More precisely, there is an automorphism σ of the graded algebra Γ such that for any graded Γ–modules M and N of type FP_∞ and for any i such that $0 \le i \le n-1$, there is a natural isomorphism of k-vector spaces

$$\varinjlim_{k \geqslant 0} \mathrm{Ext}^{n-1-i}_{\mathrm{GrMod}\,\Gamma}(J^k N, {}^\sigma M[-n]) \cong \mathrm{Hom}_{\Bbbk}\left(\varinjlim_{l \geqslant 0} \mathrm{Ext}^i_{\mathrm{GrMod}\,\Gamma}(J^l M, N), \Bbbk\right). \tag{7.1}$$

Our next goal is to establish a connection between the automorphism $\sigma : \Gamma \to \Gamma$ and the Nakayama automorphism $\nu : \Lambda \to \Lambda$ described earlier.

Theorem 9: Let Λ be an indecomposable selfinjective Koszul quiver algebra of Loewy length $n \geq 2$ with Yoneda algebra Γ. Let $F: K_\Gamma \to K_\Lambda$ be the Koszul duality, $\sigma: \Gamma \to \Gamma$ the automorphism in the Serre duality formula (7.1), and $\nu : \Lambda \to \Lambda$ the Nakayama automorphism. Then, for any Koszul Γ-module M, there is an isomorphism $\nu^{-1}F(M) \simeq F({}^{\sigma}M)$ in the category $\underline{\mathrm{GrMod}\Lambda}$ of graded Λ-modules modulo projectives.

Proof: Let N be an arbitrary Koszul Γ-module. We shall rewrite both sides of isomorphism (7.1) under the assumption that i = 0. By Theorem 6 on p. 74, the left-hand side is naturally isomorphic to

$$\underline{\mathrm{Ext}}^{-1}_{\mathrm{Gr\,Mod}\,\Lambda}\left(F\left({}^{\sigma}M\right)[-n], F(N)\right)$$

By Theorem 7 on p. 77, this is naturally isomorphic to $\mathrm{Tor}_1^{\mathrm{GrMod}\Lambda}\left(\Omega \mathrm{Tr} F({}^{\sigma}M), F(N)[n]\right)$. The right-hand side of (7.1) is naturally isomorphic, by Theorem 6 on p. 74, to $D\left(\underline{\mathrm{Ext}}^{0}_{\mathrm{GrMod}\Lambda}\left(F(N), F(M)\right)\right)$. Applying the duality D(–) to both sides, we have a natural isomorphism

$$D\left(\mathrm{Tor}_1^{\mathrm{Gr\,Mod}\,\Lambda}\left(\Omega\,\mathrm{Tr}\,F\left({}^{\sigma}M\right), F(N)[n]\right)\right) \cong \underline{\mathrm{Ext}}^{0}_{\mathrm{Gr\,Mod}\,\Lambda}\left(F(N), F(M)\right) \tag{7.2}$$

Using the graded version

$$\mathrm{Ext}^{*}_{\mathrm{Gr\,Mod}\,\Lambda^{\mathrm{op}}}\left(A, \mathrm{Hom}_{\Bbbk}(B, \Bbbk)\right) \to \mathrm{Hom}_{\Bbbk}\left(\mathrm{Tor}_{*}^{\mathrm{Gr\,Mod}\,\Lambda}(A, B), \Bbbk\right)$$

of the duality isomorphism[3, Proposition VI.5.1], the left-hand side of (7.2) is isomorphic to

$$\mathrm{Ext}^{1}_{\mathrm{Gr\,Mod}\,\Lambda^{\mathrm{op}}}\left(\Omega\,\mathrm{Tr}\,F\left({}^{\sigma}M\right), D\left[F(N)[n]\right]\right) \cong \mathrm{Ext}^{1}_{\mathrm{Gr\,Mod}\,\Lambda}\left(F(N)[n], D[\Omega\,\mathrm{Tr}\,F\left({}^{\sigma}M\right)]\right)$$

The latter is $\underline{\mathrm{Ext}}^{1}_{\mathrm{GrMod}\Lambda}\left(F(N)[n], D\left[\Omega \mathrm{Tr} F\left({}^{\sigma}M\right)\right]\right)$. Since the duality D interchanges projectives and injectives and since Λ is selfinjective, $D\Omega \mathrm{Tr} F\left({}^{\sigma}M\right) \cong \Omega^{-1} D \mathrm{Tr} F\left({}^{\sigma}M\right)$.

Notice that this is a natural isomorphism modulo projectives. Using the selfinjectivity of Λ again, the latter is naturally isomorphic modulo projectives (see, for example, [1, Proposition IV.3.7(a)] to $\Omega^{-1}\Omega^{2}D[(F({}^{\sigma}M))^{*}] = \Omega D[(F({}^{\sigma}M))^{*}]$, where $(-)^{*} = \mathrm{Homgr}_{\Lambda}(-,\Lambda)$. Combining this with a dimension shift in the second variable, we have a natural isomorphism

$$\underline{\mathrm{Ext}}^{1}_{\mathrm{Gr\,Mod}\,\Lambda}\left(F(N)[n], D[\Omega\,\mathrm{Tr}\,F({}^{\sigma}M)]\right) \cong \underline{\mathrm{Ext}}^{0}_{\mathrm{Gr\,Mod}\,\Lambda}\left(F(N)[n], D[(F({}^{\sigma}M))^{*}]\right)$$

After applying the inverse $(D(-))^{*}[n]$ of the Nakayama equivalence $D[(-)^{*}][-n]$, the latter

is naturally isomorphic to $\underline{\mathrm{Ext}}^{0}_{\mathrm{GrMod}\Lambda}\left([D(F(N))]^{*}[n], F({}^{\sigma}M)\right)$. By Lemma 14 on p. 78 and Proposition 7 on p. 79, this is naturally isomorphic to $\underline{\mathrm{Ext}}^{0}_{\mathrm{GrMod}\Lambda}\left(\nu^{-1}F(N), \rho^{-1}F(M)\right)$, where $\rho : \Lambda \to \Lambda$ is induced by $\sigma: \Gamma \to \Gamma$. As a result, having rewritten the left-hand side of (7.2), we have established a natural isomorphism

$$\underline{\mathrm{Ext}}^{0}_{\mathrm{Gr\,Mod}\,\Lambda}\left({}^{\nu^{-1}}F(N), {}^{\rho^{-1}}F(M)\right) \cong \underline{\mathrm{Ext}}^{0}_{\mathrm{Gr\,Mod}\,\Lambda}\left(F(N), F(M)\right)$$

In other words, we have a natural isomorphism

$$\underline{\mathrm{Hom}}_{\mathrm{Gr\,Mod}\,\Lambda}\left(F(N), F(M)\right) \cong \underline{\mathrm{Hom}}_{\mathrm{Gr\,Mod}\,\Lambda}\left({}^{\nu^{-1}}F(N), {}^{\rho^{-1}}F(M)\right)$$

The right-hand side can be rewritten as $\underline{\mathrm{Hom}}_{\mathrm{GrMod}\Lambda}(F(N), \nu\rho^{-1}F(M))$ and, therefore, by Yoneda's Lemma, there is an isomorphism $F(M) \simeq \nu\rho^{-1}F(M)$ in the category of Λ-modules modulo projectives, which, in view of Proposition 7 on p. 79, yields an isomorphism $\nu^{-1}F(M) \simeq F({}^{\sigma}M)$ modulo projectives. This establishes the desired connection between the Nakayama automorphism $\nu: \Lambda \to \Lambda$ and the automorphism $\sigma: \Gamma \to \Gamma$ appearing in the Serre duality formula.

NONCOMMUTATIVE SERRE DUALITY

The proof of Theorem 9 suggests a new proof of the Serre duality for generalized Artin–Schelter regular Koszul algebras (see [9, Theorem 4.9]. Let Γ be such an algebra with Yoneda algebra Λ of Loewy length, say, $n \geq 2$. Let $F: K_\Gamma \to K_\Lambda$ denote the Koszul duality. Let $\nu^{-1}:\Lambda\to\Lambda$ be the inverse of the Nakayama automorphism. By Proposition 7 on p. 79, there is an automorphism $\sigma:\Gamma \to\Gamma$ such that $F^{-1}(\nu^{-1}N)$ is naturally isomorphic, to ${}^{\sigma}F^{-1}(N)$ for any Koszul Λ-module N. After the substitution $N:= F(M)$ we have that $\nu^{-1}F(M)$ is naturally isomorphic to $F({}^{\sigma}M)$, where M runs through the category $K\Gamma$. We can now prove the Serre duality formula.

Theorem 10: Let Γ be a generalized Artin–Schelter regular Koszul quiver algebra of graded small global dimension n and $\sigma:\Gamma \to\Gamma$ the algebra automorphismdefined above. If M and N are Γ -modules of type FP_∞, then for any integer i with $0 \leq i \leq n-1$ there exists a natural isomorphism

$$D(\mathrm{Ext}^{n-1-i}_{\mathrm{Tails}\,\Gamma}(\widetilde{N}, {}^{\sigma}\widetilde{M}[-n])) \cong \mathrm{Ext}^{i}_{\mathrm{Tails}\,\Gamma}(\widetilde{M}, \widetilde{N})$$

Proof: By Theorem 2 on p. 66, we may assume, without loss of generality, that both M and N are shifts of Koszul modules. By remark (4) on p. 65, we may also assume that M and N are generated in the same degree. Therefore, shifting both modules to degree zero we may assume that both M and N are Koszul. Switching, by Theorem 6 on p. 74, from tail cohomology over Γ to Vogel cohomology over Λ and taking into account that $F({}^{\sigma}M) \cong \nu^{-1}F(M)$, we have to establish a natural isomorphism

$$D(\underline{\mathrm{Ext}}^{-i-1}_{\mathrm{Gr\,Mod}\,\Lambda}({}^{\nu^{-1}}F(M)[-n], F(N))) \cong \underline{\mathrm{Ext}}^{i}_{\mathrm{Gr\,Mod}\,\Lambda}(F(N), F(M)).$$

But this is a consequence of a more general result, which is stated next.

Proposition 8: Let Λ be a graded Frobenius algebra of Loewy length n, ν a Nakayama automorphism of Λ, and A and B finitely generated graded Λ-modules. Then, for any integer i , there is a natural isomorphism

$$D\left(\underline{\mathrm{Ext}}^{-i-1}_{\mathrm{Gr\,Mod}\,\Lambda}\left({}^{\nu^{-1}}A[-n], B\right)\right) \cong \underline{\mathrm{Ext}}^{i}_{\mathrm{Gr\,Mod}\,\Lambda}(B, A) \tag{8.1}$$

Proof: Using dimension shifts in both arguments of Vogel cohomology, we rewrite the left-hand side as

$$D\left(\underline{\mathrm{Hom}}_{\mathrm{Gr\,Mod}\,\Lambda}\left({}^{\nu^{-1}}A, \Omega^{i+1}B[n]\right)\right) \cong D\left(\mathrm{Tor}_1^{\mathrm{Gr\,Mod}\,\Lambda}\left(\mathrm{Tr}^{\nu^{-1}}A, \Omega^{i+1}B[n]\right)\right)$$

By the Tor–Ext duality, this is $\mathrm{Ext}^1_{\mathrm{GrMod}\Lambda^{\mathrm{op}}}\left(\mathrm{Tr}^{\nu^{-1}}A, D\left(\Omega^{i+1}B[n]\right)\right)$. Under the duality D, the latter is isomorphic to $\mathrm{Ext}^1_{\mathrm{GrMod}\Lambda}\left(\Omega^{i+1}B[n], \mathrm{Tr}^{\nu^{-1}}A\right)$ which is isomorphic to $\underline{\mathrm{Ext}}^1_{\mathrm{GrMod}\Lambda}\left(\Omega^{i+1}B[n], \mathrm{DTr}^{\nu^{-1}}A\right)$. Invoking the (graded) Nakayama equivalence N and using the functor isomorphism $\mathrm{DTr}(-) \cong \Omega^2\mathrm{N}(-)[n]$ (in the stable category of graded Λ-modules), we rewrite this as

$$\underline{\mathrm{Ext}}^{i+2}_{\mathrm{Gr\,Mod}\,\Lambda}\left(B[n], \Omega^2\mathcal{N}\left({}^{\nu^{-1}}A\right)[n]\right) \cong \underline{\mathrm{Ext}}^{i}_{\mathrm{Gr\,Mod}\,\Lambda}\left(B, \mathcal{N}\left({}^{\nu^{-1}}A\right)\right) \cong \underline{\mathrm{Ext}}^{i}_{\mathrm{Gr\,Mod}\,\Lambda}(B, A),$$

where the last isomorphism follows from Lemma 14 on p. 78. This finishes the proof of the proposition.

Remarks

(1) The isomorphism established in the above proposition may be called the Serre duality for Frobenius algebras.
(2) Idun Reiten pointed out to us that (8.1) is equivalent, over Frobenius algebras, to the (graded) Auslander–Reiten formula

$$D\left(\mathrm{Ext}^1_{\mathrm{Gr\,Mod}\,\Lambda}(M, D\,\mathrm{Tr}\,N)\right) \cong \underline{\mathrm{Hom}}_{\mathrm{Gr\,Mod}\,\Lambda}(N, M).$$

Toseethis,useadimensionshiftontheleft-handsideof(8.1)andrewriteitas $D\left(\underline{\mathrm{Ext}}^1_{\mathrm{GrMod}\Lambda}\left(\nu^{-1}A[-n], \Omega^{i+2}B\right)\right)$, which is $D\left(\mathrm{Ext}^1_{\mathrm{GrMod}\Lambda}\left(\nu^{-1}A[-n], \Omega^{i+2}B\right)\right)$. Applying the Nakayama automorphism to both arguments and using Lemma 14, we have $D\left(\mathrm{Ext}^1_{\mathrm{GrMod}\Lambda}\left(A[-n], \mathrm{N}\Omega^{i+2}B\right)\right)$. Replacing $\mathrm{N}\Omega^2$ by DTr[−n], we have

$D(\mathrm{Ext}^1_{\mathrm{GrMod}\,\Lambda}(\Lambda[-n], D\,\mathrm{Tr}\,\Omega^i B)[-n])$

Shifting both arguments and using the Auslander–Reiten formula,we have $\underline{\mathrm{Ext}}^1_{\mathrm{GrMod}\Lambda}(A,B)$.

REFERENCES

1. M. Auslander, I. Reiten, S. Smalø, Representation Theory of Artin Algebras, Cambridge University Press, Cambridge, 1995.
2. R.-O. Buchweitz, Tate-cohomology and maximal Cohen–Macaulay modules over Gorenstein rings, Preprint, 149 pp., 1986.
3. H. Cartan, S. Eilenberg, Homological Algebra, Princeton University Press, Princeton, 1956.
4. P. Gabriel, Des catégories abéliennes, Bull. Soc. Math. France 90 (1962) 323–448.
5. E.L. Green, R. Martínez Villa, Koszul and Yoneda algebras, I, in: Representation Theory of Algebras, in: CMS Conf. Proc., vol. 18, 1996, pp. 247–306.
6. E.L. Green, R. Martínez Villa, Koszul and Yoneda algebras, II, in: Algebras and Modules, II, in: CMS Conf. Proc., vol. 24, 1998, pp. 227–244.
7. A. Heller, Indecomposable representations and the loop-space operation, Proc. Amer. Math. Soc. 12 (1961) 640–643.
8. P. Jørgensen, Serre-duality for Tails(A), Proc. Amer. Math. Soc. 125 (1997) 709–716.
9. R. Martínez Villa, Serre duality for generalized Auslander regular algebras, Contemp. Math. 229 (1998) 237–263.
10. R. Martínez Villa, Graded, selfinjective, and Koszul algebras, J. Algebra 215 (1999) 34–72.
11. A. Martsinkovsky, New homological invariants of modules over local rings, I, J. Pure Appl. Algebra 310 (1996) 1–8.
12. J.-P. Serre, Faisceaux algébriques cohérents, Ann. of Math. 61 (1955) 197–278.
13. A.B. Verëvkin, Cohomology with noncommutative supports, Russian Math. Surveys 47 (1992) 234–235.
14. A.B. Verëvkin, Letter to the editors: Cohomology with noncommutative supports, Russian Math. Surveys 49 (1994) 255.
15. K. Yamagata, Frobenius algebras, in: M. Hazewinkel (Ed.), Handbook of Algebra, vol. 1, Elsevier, 1996, pp. 841–887.
16. A. Yekutieli, J.J. Zhang, Serre duality for noncommutative projective schemes, Proc. Amer. Math. Soc. 125 (1997) 697–707.

CITATION

1. Roberto Marti´nez Villa, Alex Martsinkovsky, Cohomology of tails, Tate–Vogel cohomology, and noncommutative Serre duality over Koszul quiver algebras, Journal of Algebra, Volume 280, Issue 1, 1 October 2004, Pages 58-83, ISSN 0021-8693, http://dx.doi.org/10.1016/j.jalgebra.2004.05.017.

Noncommutative Localization in Algebraic *L*-Theory

Andrew Ranicki
School of Mathematics, University of Edinburgh, James Clerk Maxwell Building, King's Buildings, Mayfield Road, Edinburgh EH9 3JZ, Scotland, UK

4

ABSTRACT

Given a noncommutative (Cohn) localization $A \to \sigma^{-1}A$ which is injective and stably flat we obtain a lifting theorem for induced f.g. projective $\sigma^{-1}A$-module chain complexes and localization exact sequences in algebraic L-theory, matching the algebraic K-theory localization exact sequence of Neeman–Ranicki [Amnon Neeman, Andrew Ranicki, Noncommutative localisation in algebraic K-theory I, Geom. Topol. 8 (2004) 1385–1425] and Neeman [Amnon Neeman, Noncommutative localisation in algebraic K-theory II, Adv. Math. 213 (2007) 785–819].

INTRODUCTION

The series of papers [2,3], studied the algebraic K-theory of the noncommutative (Cohn) localization σ^{-1} A of a ring A inverting a collection σ of morphisms of f.g. projective left Amodules. By definition, σ^{-1} A is stably flat if

$$\mathrm{Tor}_i^A(\sigma^{-1}A, \sigma^{-1}A) = 0 \quad (i \geqslant 1).$$

An (A, σ)-module is an A-module T which admits an f.g. projective A-module resolution

$$0 \longrightarrow P \stackrel{s}{\longrightarrow} Q \longrightarrow T \longrightarrow 0$$

with $s : \sigma^{-1} P \to \sigma^{-1} Q$ an isomorphism of the induced σ^{-1} A-modules. For $A \to \sigma^{-1} A$ which is injective and stably flat we obtained an algebraic K-theory localization exact sequence

$$\cdots \longrightarrow K_n(A) \longrightarrow K_n(\sigma^{-1}A) \longrightarrow K_{n-1}(H(A,\sigma)) \longrightarrow K_{n-1}(A) \longrightarrow \cdots$$

with $H(A, \sigma)$ the exact category of (A, σ)-modules. Let C be a bounded σ^{-1} A-module chain complex such that each $C_i = \sigma^{-1}P_i$ is induced from an f.g. projective A-module P_i. The chain complex lifting problem is to decide if C is chain equivalent to $\sigma^{-1}D$ for a bounded chain complex D of f.g. projective A-modules. The problem has a trivial affirmative solution for a commutative or Ore localization, by the clearing of denominators, when C is actually isomorphic to $\sigma^{-1}D$. In general, it is not possible to lift chain complexes: the injective noncommutative localizations $A \to \sigma^{-1}A$ which are not stably flat constructed in Neeman, Ranicki and Schofield [4, Remark 2.13] provide examples of induced f.g. projective σ^{-1} A-module chain complexes of dimensions 3 which cannot be lifted. In Section 1 we solve the chain complex lifting problem in the injective stably flat case, obtaining the following results (Theorems 1.4, 1.5):

Theorem 0.1: For a stably flat injective noncommutative localization $A \to \sigma^{-1}A$ every bounded chain complex C of induced f.g. projective σ^{-1}A-modules is chain equivalent to $\sigma^{-1}D$ for a bounded chain complex D of f.g. projective A-modules. Moreover, if C is n-dimensional

$$C : \cdots \longrightarrow 0 \longrightarrow C_n \longrightarrow C_{n-1} \longrightarrow \cdots \longrightarrow C_1 \longrightarrow C_0 \longrightarrow 0 \longrightarrow \cdots$$

then D can be chosen to be n-dimensional.

In Section 2 we consider the algebraic L-theory of a noncommutative localization, obtaining the following results (Theorems 2.4, 2.5, 2.9):

Theorem 0.2: Let $A \to \sigma^{-1}A$ be a noncommutative localization of a ring with involution A, such that σ is invariant under the involution.

i. There is a localization exact sequence of quadratic L-groups

$$\cdots \longrightarrow L_n(A) \longrightarrow L_n^I(\sigma^{-1}A) \xrightarrow{\partial} L_n(A,\sigma) \longrightarrow L_{n-1}(A) \longrightarrow$$

with $I = \text{im}(K_0(A) \to K_0(\sigma^{-1}A))$, and $L_n(A, \sigma)$ the cobordism group of $\sigma^{-1}A$ contractible $(n-1)$-dimensional quadratic Poincaré complexes over A.

ii. If $\sigma^{-1}A$ is stably flat over A there is a localization exact sequence of symmetric L-groups

$$\cdots \longrightarrow L^n(A) \longrightarrow L_I^n(\sigma^{-1}A) \xrightarrow{\partial} L^n(A,\sigma) \longrightarrow L^{n-1}(A) \longrightarrow$$

with $L^n(A, \sigma)$ the cobordism group of $\sigma^{-1}A$-contractible $(n-1)$-dimensional symmetric Poincaré complexes over A.

iii. If $A \to \sigma^{-1}A$ is injective then $L^n(A, \sigma)$ (resp. $L^n(A, \sigma)$) is the cobordism group of ndimensional symmetric (resp. quadratic) Poincaré complexes of (A, σ)-modules.

The L-theory exact sequences of Theorem 0.2 for an injective Ore localization $A \to \sigma^{-1}A$ (which is flat and hence stably flat) were obtained in Ranicki [5]. The quadratic L-theory exact sequence of 0.2(i) for arbitrary injective $A \to \sigma^{-1}A$ was obtained by Vogel [8, 9]. The symmetric L-theory exact sequence of 0.2(ii) is new. We refer to [6, 7] for some of the applications of the algebraic L-theory of noncommutative localizations to topology. Amnon Neeman used to be a coauthor of the paper, but decided to withdraw in May 2007.

LIFTING CHAIN COMPLEXES

If $A \to \sigma^{-1}A$ is a stably flat localization, we know from [3, Theorem 0.4, Proposition 4.5 and Theorem 3.7] that the functor $\text{Ti}: \frac{D^{\text{perf}}}{R^c} \to D^{\text{perf}}(\sigma^{-1}A)$ is just an idempotent completion; it is fully faithful and all objects in $D^{\text{perf}}(\sigma^{-1}A)$ are, up to isomorphisms, direct summands of objects in the image of Ti. A fairly easy consequence of this is the following. Let $C \in D^{perf}(\sigma^{-1}A)$ be the complex

$$0 \longrightarrow \sigma^{-1}C^m \longrightarrow \sigma^{-1}C^{m+1} \longrightarrow \cdots \longrightarrow \sigma^{-1}C^{n-1} \longrightarrow \sigma^{-1}C^n \longrightarrow 0,$$

with C^i all finitely generated, projective A-modules. Then there is complex $X \in D^{perf}(A)$ with $C \simeq \{\sigma^{-1}A\}^L \otimes_A X$. That is, C is homotopy equivalent to the tensor product with $\sigma^{-1}A$ of a perfect complex over the ring A. In Section 1 we prove this (Theorem 1.4), and then refine the result to show that X may be chosen to be a complex of the form

$$0 \longrightarrow X^m \longrightarrow X^{m+1} \longrightarrow \cdots \longrightarrow X^{n-1} \longrightarrow X^n \longrightarrow 0.$$

(Proof in Theorem 1.5.)

Remark 1.1: The proof of Theorem 1.4 relies on the following fact about triangulated categories. Suppose A is a full, triangulated subcategory of a triangulated category B, and suppose all objects in B are direct summands of objects of A. An object $X \in B$ belongs to $A \subset B$ if and only if $[X] \in K_0(B)$ lies in the image of $K_0(A) \to K_0(B)$. This fact may be found, for example, in [1, Proposition 4.5.11], but for the reader's convenience its proof is included here in Lemma 1.2 and Proposition 1.3.

We begin by reminding the reader of some basic facts about Grothendieck groups. For any additive category A we define K_0^{add} (A) to be the Grothendieck group of the split exact category A. This means that the short exact sequences in A are precisely the split sequences. It is well known that every element of K_0^{add} (A) can be expressed as

$$[X]-[Y]$$

for X and Y objects of A. The expressions $[X]-[Y]$ and $[X']-[Y']$ are equal in K_0^{add} (A) if and only if there exists an object $P \in A$ and an isomorphism

$$X \oplus Y' \oplus P = X' \oplus Y \oplus P.$$

If A happens to be a triangulated category, then $K_0(A)$ means the quotient of K_0^{add} (A) by a certain subgroup we will denote T (A). The subgroup T (A) is defined as the group generated by all

$$[X] - [Y] + [Z],$$

where there exists a distinguished triangle in A

$$X \longrightarrow Y \longrightarrow Z \longrightarrow \Sigma X.$$

We prove:

Lemma 1.2: Suppose B is a triangulated category. Let A be a full, triangulated subcategory of B. Assume further that every object of B is a direct summand of an object in A ⊂ B. Then the map f: K_0^{add} (A) → K_0^{add} (B) induces a surjection T (A) → T (B). In symbols: f (T (A)) = T (B).

Proof: Let [X]−[Y]+[Z] be a generator of T (B) ⊂ K_0^{add} (B). We need to show it lies in the image of T (A) ⊂ K_0^{add} (A). Suppose therefore that

$$X \longrightarrow Y \longrightarrow Z \longrightarrow \Sigma X$$

is a distinguished triangle in B. Because every object of B is a direct summand of an object in A, we can choose objects C and D with

$$X \oplus C, \qquad Z \oplus D$$

both lying in A. But then we have two distinguished triangles in B

$$X \longrightarrow Y \longrightarrow Z \longrightarrow \Sigma X,$$

$$C \longrightarrow C \oplus D \longrightarrow D \xrightarrow{0} \Sigma C$$

and their direct sum is a distinguished triangle

$$X \oplus C \longrightarrow Y \oplus C \oplus D \longrightarrow Z \oplus D \longrightarrow \Sigma(X \oplus C).$$

Two of the objects lie in A. Since the subcategory A ⊂ B is full and triangulated, the entire distinguished triangle lies in A. Thus

$$[X \oplus C] - [Y \oplus C \oplus D] + [Z \oplus D] = [X] - [Y] + [Z]$$

lies in the image of T (A).

The next proposition is well known; again, the proof is included for the convenience of the reader.

Proposition 1.3: Let the hypotheses be as in Lemma 1.2. That is, suppose B is a triangulated category. Let A be a full, triangulated subcategory of B. Assume further that every object of B is a direct summand of an object in A ⊂ B. If X is an object of B and [X] lies in the image of the natural map f: $K_0(A) \dashrightarrow K_0(B)$, then X ∈ A.

Proof: If we consider [X] as an element of K_0^{add} (B), then saying that its image in $K_0(B)$ lies in the image of $K_0(A) \rightarrow K_0(B)$ is equivalent to saying that, modulo T (B), [X] lies in the image of K_0^{add} (A). That is,

$$[X] \in T(\mathcal{B}) + f\left(K_0^{\text{add}}(\mathcal{A})\right) \subset K_0^{\text{add}}(\mathcal{B}).$$

By Lemma 1.2 we have that f (T (A)) = T (B). Thus

$$\begin{aligned} T(\mathcal{B}) + f\left(K_0^{\text{add}}(\mathcal{A})\right) &= f\left(T(\mathcal{A})\right) + f\left(K_0^{\text{add}}(\mathcal{A})\right) \\ &= f\left(K_0^{\text{add}}(\mathcal{A})\right). \end{aligned}$$

That means there exist objects C and D in A ⊂ B and an identity in K_0^{add} (B)

$$[X] = [C] - [D].$$

There must therefore be an object P ∈ B and an isomorphism

$$X \oplus D \oplus P \simeq C \oplus P.$$

But P is an object of B, hence a direct summand of an object of A. There is an object P' ∈ B with P ⊕ P' ∈ A. We have an isomorphism

$$X \oplus D \oplus P \oplus P' \simeq C \oplus P \oplus P'.$$

Putting D′ = D ⊕ P ⊕ P and C′ = C ⊕ P ⊕ P′ we have objects C′, D′ in A, and a (split) distinguished triangle

$$D' \longrightarrow C' \longrightarrow X \longrightarrow \Sigma D'.$$

Since A ⊂ B is triangulated we conclude that X ∈ A.

The relevance of these results to our work here is

Theorem 1.4: Let A → σ^{-1}A be a stably flat localization of rings. Suppose we are given a perfect complex C over σ^{-1}A. Suppose further that C ∈ $D^{perf}(\sigma^{-1}A)$ is of the form

$$0 \longrightarrow \sigma^{-1}C^m \longrightarrow \sigma^{-1}C^{m+1} \longrightarrow \cdots \longrightarrow \sigma^{-1}C^{n-1} \longrightarrow \sigma^{-1}C^n \longrightarrow 0$$

where each C^i is a finitely generated, projective A-module. Then C is homotopy equivalent to $\{\sigma^{-1}A\}^L \otimes_A X$, for some X ∈ $D^{perf}(A)$.

Proof: The localization is stably flat. By [3, Theorem 0.4] the functor T: $T^c \to D^{perf}(\sigma^{-1}A)$ is an equivalence of categories. By [3, Proposition

4.5 and Theorem 3.7] we also know that the functor $i : \frac{D^{perf}(A)}{R^c} \to T^c$ is fully faithful, and that every object in T^c is isomorphic to a direct summand of an object in the image of i. Next we apply Proposition 1.3, with B = $D^{perf}(\sigma^{-1}A)$ and A the full subcategory containing all objects isomorphic to $T_i(x)$, for any $x \in \frac{D^{perf}(A)}{R^c}$. Now C is an object of $D^{perf}(\sigma^{-1}A)$, and in $K_0(D^{perf}(\sigma^{-1}A))$ we have an identity

$$[C] = \sum_{\ell=-\infty}^{\infty} (-1)^{\ell} [\sigma^{-1} C^{\ell}]$$

with

$$[\sigma^{-1} C^{\ell}] = [\{\sigma^{-1} A\} \otimes_A C^{\ell}] = [TiC^{\ell}]$$

certainly lying in the image of the map

$$K_0(Ti) : K_0\left(\frac{D^{perf}(A)}{\mathcal{R}^c}\right) \longrightarrow K_0(D^{perf}(\sigma^{-1}A)).$$

Proposition 1.3: therefore tells us that C is isomorphic to an object in the image of the functor Ti. There exists a perfect complex X ∈ $D^{perf}(A)$ and a homotopy equivalence $C \simeq \{\sigma^{-1}A\}^L \otimes_A X$.

The problem with Theorem 1.4 is that it gives us no bound on the length of the complex X with $\{\sigma^{-1}A\}^L \otimes_A X \simeq C$. We really want to know

Theorem 1.5: Let $A \to \sigma^{-1}A$ be a stably flat localization of rings. Suppose C ∈ $D^{perf}(\sigma^{-1}A)$ is the complex

$$0 \longrightarrow \sigma^{-1}C^m \longrightarrow \sigma^{-1}C^{m+1} \longrightarrow \cdots \longrightarrow \sigma^{-1}C^{n-1} \longrightarrow \sigma^{-1}C^n \longrightarrow 0.$$

Then the complex X ∈ $D^{perf}(A)$ with $C \simeq \{\sigma^{-1}A\}^L \otimes_A X$, whose existence is guaranteed by Theorem 1.4, may be chosen to be a complex

$$0 \longrightarrow X^m \longrightarrow X^{m+1} \longrightarrow \cdots \longrightarrow X^{n-1} \longrightarrow X^n \longrightarrow 0.$$

If m = n this is easy. For m<n we need to prove something. Our proof will appeal to the results of [3, Section 4]. We remind the reader that this was the section which dealt with the subcategories K[m, n] of complexes in R^c vanishing outside the range [m, n]. First we need a lemma.

Lemma 1.6: Let M and N be any finitely generated projective A-modules. We may view M and N as objects in the derived category $D^{perf}(A)$, concentrated in degree 0. Then any map in $T^c(\pi M, \pi N)$ can be represented as $\pi(\alpha)^{-1}\pi(\beta)$, for some α, β morphisms in $D^{perf}(A)$ as below

$$M \xrightarrow{\beta} Y \xleftarrow{\alpha} N.$$

The map $\alpha: N \to Y$ fits in a triangle

$$X \longrightarrow N \xrightarrow{\alpha} Y \longrightarrow \Sigma X$$

and X may be chosen to lie in K[0, 1].

Proof: By [3, Proposition 4.5 and Theorem 3.7] we know that the map

$$i : \frac{D^{\text{perf}}(A)}{\mathcal{R}^c} \longrightarrow \mathcal{T}^c$$

is fully faithful. Therefore

$$\mathcal{T}^c(\pi M, \pi N) = \frac{D^{\text{perf}}(A)}{\mathcal{R}^c}(M, N).$$

That is, any map $\pi M \to \pi N$ can be written as $\pi(\alpha)^{-1}\pi(\beta)$, for some α, β morphisms in $D^{perf}(A)$ as below

$$M \xrightarrow{\beta} Y \xleftarrow{\alpha} N.$$

The map α: N → Y fits in a triangle

$$X \longrightarrow N \xrightarrow{\alpha} Y \xrightarrow{\beta} \Sigma X$$

and X may be chosen to lie in R^c. What is not clear is that we may choose X in K[0, 1] ⊂ R^c. The easy observation is that we may certainly modify our choice of X to lie in K ⊂ R^c. This follows from [2, Lemma 4.5], which tells us that for any choice of X as above there exists an X′ with X ⊕ X′ isomorphic to an object in K. We have a distinguished triangle

$$X \oplus X' \longrightarrow N \xrightarrow{\binom{\alpha}{0}} Y \oplus \Sigma X' \xrightarrow{\beta \oplus 1} \Sigma(X \oplus X')$$

and a diagram

$$M \xrightarrow{\binom{\beta}{0}} Y \oplus \Sigma X' \xleftarrow{\binom{\alpha}{0}} N,$$

and replacing our original choices by these we may assume X ∈ K. Now we have to shorten X. By [2, Lemma 4.7], there exists a triangle in R^c

$$X' \longrightarrow X \longrightarrow X'' \longrightarrow \Sigma X'$$

with X′ ∈ K[1,∞) and X″ ∈ K(−∞, 1]. The composite X′ → X → N is a map from X′ ∈ K[1,∞) to N ∈ $S^{\leq 0}$, which must vanish. Hence we have that X → N factors as X → X′ → N. We complete to a morphism of triangles

$$\begin{array}{ccccccc} X & \longrightarrow & N & \xrightarrow{\alpha} & Y & \longrightarrow & \Sigma X \\ \downarrow & & \downarrow{\scriptstyle 1} & & \downarrow{\scriptstyle \gamma} & & \downarrow \\ X'' & \longrightarrow & N & \xrightarrow{\gamma\alpha} & Y'' & \longrightarrow & \Sigma X'' \end{array}$$

and another representative of our morphism is the diagram

$$M \xrightarrow{\gamma\beta} Y'' \xleftarrow{\gamma\alpha} N.$$

We may, on replacing Y by Y″, assume $X \in K(-\infty, 1]$.

Applying [2, Lemma 4.7] again, we have that any $X \in K(-\infty, 1]$ admits a triangle

$$X' \longrightarrow X \longrightarrow X'' \longrightarrow \Sigma X'$$

with $X' \in K[0, 1]$ and $X'' \in K(-\infty, 0]$. Form the octahedron

$$\begin{array}{ccccccc}
X' & \longrightarrow & N & \xrightarrow{\alpha'} & Y' & \longrightarrow & \Sigma X' \\
\downarrow & & \downarrow{\scriptstyle 1} & & \downarrow{\scriptstyle \gamma} & & \downarrow \\
X & \longrightarrow & N & \xrightarrow{\alpha} & Y & \longrightarrow & \Sigma X \\
 & & & & \downarrow & & \downarrow \\
 & & & & \Sigma X'' & \xrightarrow{1} & \Sigma X''.
\end{array}$$

The composite $M \to Y \to \Sigma X''$ is a map from the projective module M, viewed as a complex concentrated in degree 0, to $\Sigma X'' \in K(\infty, -1]$. This composite must vanish. The map $\beta: M \to Y$ therefore factors as $M \xrightarrow{\beta'} Y' \xrightarrow{\gamma} Y$ and our morphism in T^c has a representative

$$M \xrightarrow{\beta'} Y' \xleftarrow{\alpha'} N$$

so that in the triangle

$$X' \longrightarrow N \xrightarrow{\alpha'} Y' \longrightarrow \Sigma X'$$

X' may be chosen to lie in $K[0, 1]$.

Now we are ready for

Proof of Theorem 1.5: We are given a complex $C \in D^{perf}(\sigma^{-1}A)$ of the form

$$0 \longrightarrow \sigma^{-1}C^m \longrightarrow \sigma^{-1}C^{m+1} \longrightarrow \cdots \longrightarrow \sigma^{-1}C^{n-1} \longrightarrow \sigma^{-1}C^n \longrightarrow 0.$$

To eliminate the trivial case, assume $m \leq n + 1$. Shifting, we may assume $m = 0$ and $n \geq 1$. Theorem 1.4 guarantees that C is homotopy equivalent to $\{\sigma^{-1}A\}^L\otimes_A D$, with $D \in D^{perf}(A)$. But D need not be supported on the interval $[0, n]$. We need to show how to shorten D. Assume therefore that D is supported on $[-1, n]$. We will show how to replace D by a complex supported on $[0, n]$. Shortening a complex supported on $[0, n + 1]$ is dual, and we leave it to the reader. We may suppose therefore that $D \in D^{perf}(A)$ is the complex

$$\cdots \longrightarrow 0 \longrightarrow D^{-1} \longrightarrow D^0 \longrightarrow \cdots \longrightarrow D^n \longrightarrow 0 \longrightarrow \cdots$$

and that there is a homotopy equivalence of $\sigma^{-1}D$ with a shorter complex, that is a commutative diagram

$$\begin{array}{ccccccccccc}
\longrightarrow 0 & \longrightarrow & \sigma^{-1}D^{-1} & \xrightarrow{\partial} & \sigma^{-1}D^0 & \longrightarrow & \cdots & \longrightarrow & \sigma^{-1}D^n & \longrightarrow & 0 \longrightarrow \\
\downarrow & & \downarrow & & \downarrow & & & & \downarrow & & \downarrow \\
\longrightarrow 0 & \longrightarrow & 0 & \longrightarrow & \sigma^{-1}C^0 & \longrightarrow & \cdots & \longrightarrow & \sigma^{-1}C^n & \longrightarrow & 0 \longrightarrow \\
\downarrow & & \downarrow & & \downarrow & & & & \downarrow & & \downarrow \\
\longrightarrow 0 & \longrightarrow & \sigma^{-1}D^{-1} & \xrightarrow{\partial} & \sigma^{-1}D^0 & \longrightarrow & \cdots & \longrightarrow & \sigma^{-1}D^n & \longrightarrow & 0 \longrightarrow
\end{array}$$

so that the composite is homotopic to the identity. In particular, there is a map $d: \sigma^{-1}D^0 \to \sigma^{-1}D^{-1}$ so that $d\partial: \sigma^{-1}D^{-1} \to \sigma^{-1}D^{-1}$ is the identity. By [2, Proposition 3.1] the map $d: \sigma^{-1}D^0 \to \sigma^{-1}D^{-1}$ lifts uniquely to a map $d': \pi D^0 \to \pi D^{-1}$. By Lemma 1.6 the map d' can be represented as $\pi(\alpha)^{-1}\pi(\beta)$, where α and β are, respectively, the chain maps

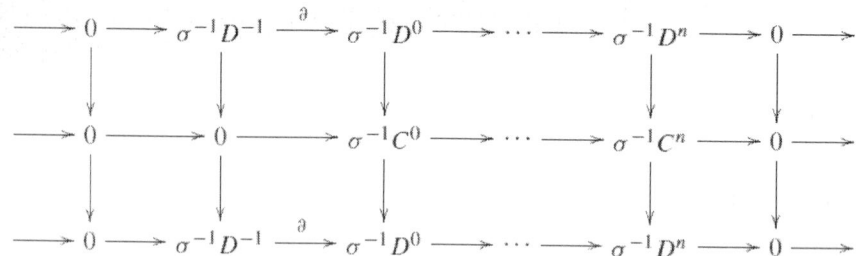

$$\begin{array}{ccccccccc}
\longrightarrow & 0 & \longrightarrow & 0 & \longrightarrow & D^{-1} & \longrightarrow & 0 & \longrightarrow \\
 & \downarrow & & \downarrow & & \downarrow & & \downarrow & \\
\longrightarrow & 0 & \longrightarrow & X & \xrightarrow{r} & Y & \longrightarrow & 0 & \longrightarrow
\end{array}$$

and

$$\begin{array}{ccccccccc}
\longrightarrow & 0 & \longrightarrow & 0 & \longrightarrow & D^{0} & \longrightarrow & 0 & \longrightarrow \\
 & \downarrow & & \downarrow & & \downarrow{\scriptstyle g} & & \downarrow & \\
\longrightarrow & 0 & \longrightarrow & X & \xrightarrow{r} & Y & \longrightarrow & 0 & \longrightarrow
\end{array}.$$

The fact that $\sigma^{-1}\alpha$ is an equivalence tells us that the map $\sigma^{-1}r$: $\sigma^{-1}X \rightarrow \sigma^{-1}Y$ is injective, with cokernel $\sigma^{-1}D^{-1}$. The fact that $\alpha^{-1}\beta$ agrees with d′ means that the composite

$$\sigma^{-1}D^0 \xrightarrow{\sigma^{-1}g} \sigma^{-1}Y \longrightarrow \mathrm{Coker}(\sigma^{-1}r)$$

is just the map d: $^{-1}$D0 → $\sigma^{-1}D^{-1}$. Let X be the chain complex

$$\longrightarrow 0 \longrightarrow D^0 \oplus X \xrightarrow{\begin{pmatrix} \partial & 0 \\ g & r \end{pmatrix}} D^1 \oplus Y \longrightarrow \cdots \longrightarrow D^n \longrightarrow 0 \longrightarrow$$

Let f: X → D be the natural map of chain complexes

$$\begin{array}{ccccccccccccccc}
\longrightarrow & 0 & \longrightarrow & 0 & \longrightarrow & D^0 \oplus X & \xrightarrow{\begin{pmatrix} \partial & 0 \\ g & r \end{pmatrix}} & D^1 \oplus Y & \longrightarrow & \cdots & \longrightarrow & D^n & \longrightarrow & 0 & \longrightarrow \\
 & & & \downarrow & & \downarrow{\scriptstyle \pi_1} & & \downarrow{\scriptstyle \pi_1} & & & & & & \downarrow & \\
\longrightarrow & 0 & \longrightarrow & D^{-1} & \longrightarrow & D^0 & \xrightarrow[\partial]{} & D^1 & \longrightarrow & \cdots & \longrightarrow & D^n & \longrightarrow & 0 & \longrightarrow
\end{array}$$

where the vertical maps labelled π_1 are the projections to the first factor of the direct sum. The map $\sigma^{-1}f$ is easily seen to be homotopy equivalence. Thus $\sigma^{-1}X$ is homotopy equivalent to $\sigma^{-1}D \cong C$.

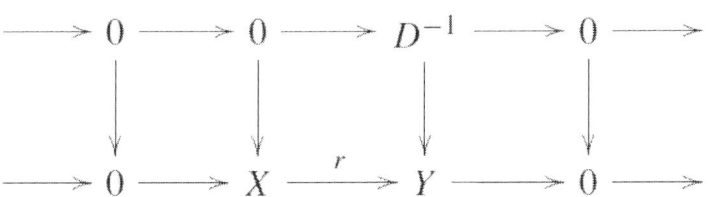

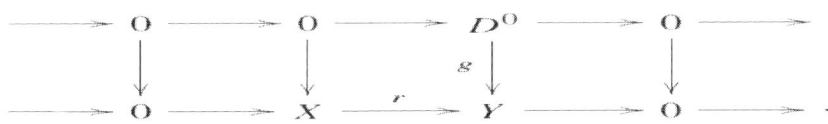

ALGEBRAIC L-THEORY

An involution on a ring A is an anti-automorphism

$$A \longrightarrow A; \quad r \longmapsto \bar{r}.$$

The involution is used to regard a left A-module M as a right A-module by

$$M \times A \longrightarrow M; \quad (x, r) \longmapsto \bar{r}x.$$

The dual of a (left) A-module M is the A-module

$$M^* = \mathrm{Hom}_A(M, A), \qquad A \times M^* \longrightarrow M^*; \quad (r, f) \longmapsto (x \longmapsto f(x)\bar{r}).$$

The dual of an A-module morphism s: P → Q is the A-module morphism

$$s^* : Q^* \longrightarrow P^*; \quad f \longmapsto (x \longmapsto f(s(x))).$$

If M is f.g. projective then so is M∗, and

$$M \longrightarrow M^{**}; \quad x \longmapsto (f \longmapsto \overline{f(x)})$$

is an isomorphism which is used to identify $M^{**} = M$.

Hypothesis 2.1: In this section, we assume that

(i) A is a ring with involution,
(ii) the duals of morphisms s : P → Q in σ are morphisms $s^*: Q^* \to P^*$ in σ,
(iii) ε ∈ A is a central unit such that $\varepsilon = \in^{-1}$ (e.g. ε = ±1).

The noncommutative localization $\sigma^{-1}A$ is then also a ring with involution, with ∈ $\sigma^{-1}A$ a central unit such that $\bar{\varepsilon} = \in^{-1}$.

We review briefly the chain complex construction of the f.g. projective -quadratic L-groups $L_*(A,)$ and the -symmetric L-groups $L^*(A,)$. Given an A-module chain complex C let the generator $T \in \mathbb{Z}_2$ act on the $\mathbb{Z}$-module chain complex $C \otimes_A C$ by the ε-transposition duality

$$T_\epsilon : C_p \otimes_A C_q \longrightarrow C_q \otimes_A C_p : x \otimes y \longmapsto (-1)^{pq} \epsilon y \otimes x .$$

Let W be the standard free $\mathbb{Z}[\mathbb{Z}_2]$-module resolution of $\mathbb{Z}$

The ε-symmetric (resp. ε-quadratic) Q-groups of C are the $\mathbb{Z}_2$-hyper-cohomology (resp. $\mathbb{Z}_2$- hyperhomology) groups of $C \otimes_A C$

$$Q^n(C, \epsilon) = H^n(\mathbb{Z}_2; C \otimes_A C) = H_n(\mathrm{Hom}_{\mathbb{Z}[\mathbb{Z}_2]}(W, C \otimes_A C)),$$
$$Q_n(C, \epsilon) = H_n(\mathbb{Z}_2; C \otimes_A C) = H_n(W \otimes_{\mathbb{Z}[\mathbb{Z}_2]} (C \otimes_A C)).$$

The Q-groups are chain homotopy invariants of C. There are defined forgetful maps

$$1 + T_\epsilon : Q_n(C, \epsilon) \longrightarrow Q^n(C, \epsilon); \quad \psi \longmapsto (1 + T_\epsilon)\psi,$$
$$Q^n(C, \epsilon) \longrightarrow H_n(C \otimes_A C); \quad \phi \longmapsto \phi_0 .$$

For f.g. projective C the function

$$C \otimes_A C \longrightarrow \mathrm{Hom}_A(C^*, C); \quad x \otimes y \longmapsto \left(f \longmapsto \overline{f(x)}y\right)$$

is an isomorphism of $\mathbb{Z}[\mathbb{Z}_2]$-module chain complexes, with $T \in \mathbb{Z}_2$ acting on $\mathrm{Hom}_A(C^*,C)$ by $\theta \mapsto \varepsilon\theta^*$. The element $\varphi_0 \in H_n(C \otimes A\, C) = H_n(\mathrm{Hom}_A(C^*, C))$ is a chain homotopy class of A-module chain maps $\varphi_0: C^{n-*} \to C$. An n-dimensional ε-symmetric complex over A (C, φ) is a bounded f.g. projective A-module chain complex C together with an element $\varphi \in Q^n(C, \varepsilon)$. The complex (C, φ) is Poincaré if the A-module chain map $\varphi_0: C^{n-*} \to C$ is a chain equivalence.

Example 2.2: A 0-dimensional ε-symmetric Poincaré complex (C, φ) over A is essentially the same as a nonsingular ε-symmetric form (M, λ) over (A, σ), with M = $(C_0)^*$ an f.g. projective A-module and

$$\lambda = \phi_0 : M \times M \longrightarrow A$$

a sesquilinear pairing such that the adjoint

$$M \longrightarrow M^*; \quad x \longmapsto (y \longmapsto \lambda(x, y))$$

is an A-module isomorphism.

See pp. 210–211 of [6] for the notion of an ε-symmetric (Poincaré) pair. The boundary of an n-dimensional ε-symmetric complex (C, φ) is the (n−1)-dimensional ε-symmetric Poincaré complex

$$\partial(C, \phi) = (\partial C, \partial \phi)$$

with ∂C = $C(\varphi_0: C^{n-*} \to C)_{*+1}$ and ∂φ as defined on p. 218 of [6]. The n-dimensional ε- symmetric L-group L^n(A,) is the cobordism group of n-dimensional ε-symmetric Poincaré complexes (C, φ) over A with C n-dimensional. In particular, L^0(A,) is the Witt group of nonsingular ε -symmetric forms over A.

An n-dimensional ε-symmetric complex (C, φ) over A is σ^{-1}A-Poincaré if the σ^{-1}Amodule chain map $\sigma^{-1}\varphi 0$: $\sigma^{-1}C^{n-*} \to \sigma^{-1}C$ is a chain equivalence, in which case σ^{-1}(C, φ) is an n-dimensional ε-symmetric Poincaré complex over σ^{-1}A.

The n-dimensional -symmetric Γ -group $\Gamma^n(A \to \sigma^{-1}A, \varepsilon)$ is the cobordism group of ndimensional ε-symmetric σ^{-1}A-Poincaré complexes (C, φ) over A such that σ^{-1}C is chain equivalent to an n-dimensional induced f.g. projective σ^{-1}A-module chain complex. The ndimensional ε-symmetric L-group $L^n(A, \sigma, \varepsilon)$ is the cobordism group of (n − 1)-dimensional ε-symmetric Poincaré complexes over A (C, φ) such that C is σ −1A-contractible, i.e. $\sigma^{-1}C \simeq 0$. Similarly in the -quadratic case, with groups $L^n(A, \varepsilon)$, $\Gamma n(A \to \sigma^{-1}A, \varepsilon)$, $L^n(A, \sigma, \varepsilon)$. The ε-quadratic L- and Γ -groups are 4-periodic

$$L_n(A,\epsilon) = L_{n+2}(A,-\epsilon) = L_{n+4}(A,\epsilon),$$
$$\Gamma_n(A \longrightarrow \sigma^{-1}A,\epsilon) = \Gamma_{n+2}(A \longrightarrow \sigma^{-1}A,-\epsilon) = \Gamma_{n+4}(A \longrightarrow \sigma^{-1}A,\epsilon),$$
$$L_n(A,\sigma,\epsilon) = L_{n+2}(A,\sigma,-\epsilon) = L_{n+4}(A,\sigma,\epsilon).$$

Proposition 2.3: For any ring with involution A and noncommutative localization $\sigma^{-1}A$ there is defined a localization exact sequence of ε -symmetric L-groups

$$\cdots \longrightarrow L^n(A,\epsilon) \longrightarrow \Gamma^n(A \longrightarrow \sigma^{-1}A,\epsilon) \xrightarrow{\partial} L^n(A,\sigma,\epsilon) \longrightarrow L^{n-1}(A,\epsilon) \longrightarrow \cdots.$$

Similarly in the -quadratic case, with an exact sequence

$$\cdots \longrightarrow L_n(A,\epsilon) \longrightarrow \Gamma_n(A \longrightarrow \sigma^{-1}A,\epsilon) \xrightarrow{\partial} L_n(A,\sigma,\epsilon) \longrightarrow L_{n-1}(A,\epsilon) \longrightarrow \cdots.$$

Proof: The relative group of $L^n(A, \varepsilon) \dashrightarrow \Gamma^n(A \dashrightarrow \sigma^{-1}A, \varepsilon)$ is the cobordism group of n-dimensional ε-symmetric $\sigma^{-1}A$-Poincaré pairs over A $(f\colon C \to D, (\delta\varphi, \varphi))$ with (C, φ) Poincaré. The effect of algebraic surgery on (C, φ) using this pair is a cobordant (n − 1)- dimensional ε-symmetric Poincaré complex (C', φ') with C' $\sigma^{-1}A$-contractible. The function

$(f\colon C\to D,(\delta\varphi,\varphi)) \to (C', \varphi')$ defines an isomorphism between the relative group and $L^n(A, \sigma, \varepsilon)$.

Define

$$I = \mathrm{im}(K_0(A) \longrightarrow K_0(\sigma^{-1}A)),$$

the subgroup of $K_0(\sigma^{-1}A)$ consisting of the projective classes of the f.g. projective $\sigma^{-1}A$modules induced from f.g. projective A-modules. By definition, $L^n_I(\sigma^{-1}A,\varepsilon)$ is the cobordism group of n-dimensional ε-symmetric Poincaré complexes over $\sigma^{-1}A(B, \theta)$ such that $[B] \in I$. There are evident morphisms of Γ - and L-groups

$$\sigma^{-1}\Gamma^* : \Gamma^n(A \longrightarrow \sigma^{-1}A,\epsilon) \longrightarrow L^n_I(\sigma^{-1}A,\epsilon); \quad (C,\phi) \longmapsto \sigma^{-1}(C,\phi),$$
$$\sigma^{-1}\Gamma_* : \Gamma_n(A \longrightarrow \sigma^{-1}A,\epsilon) \longrightarrow L^I_n(\sigma^{-1}A,\epsilon); \quad (C,\psi) \longmapsto \sigma^{-1}(C,\psi).$$

In general, the morphisms $\sigma^{-1}\Gamma^*$, $\sigma^{-1}\Gamma^*$ need not be isomorphisms, since a bounded f.g. projective $\sigma^{-1}A$-module chain complex D with $[D] \in I$ need not be chain equivalent to $\sigma^{-1}C$ for a bounded f.g. projective A-module chain complex C. It was proved in Chapter 3 of Ranicki [5] that if $A \to \sigma^{-1}A$ is an injective Ore localization then the morphisms $\sigma^{-1}Q^*$, $\sigma^{-1}Q^*$, $\sigma^{-1}\Gamma_*$, $\sigma^{-1}\Gamma_*$ are isomorphisms, so that there are defined localization exact sequences for both the -symmetric and the ε-quadratic L-groups

$$\cdots \longrightarrow L^n(A,\epsilon) \longrightarrow L^n_I(\sigma^{-1}A,\epsilon) \xrightarrow{\partial} L^n(A,\sigma,\epsilon) \longrightarrow L^{n-1}(A,\epsilon) \longrightarrow \cdots,$$

$$\cdots \longrightarrow L_n(A,\epsilon) \longrightarrow L^I_n(\sigma^{-1}A,\epsilon) \xrightarrow{\partial} L_n(A,\sigma,\epsilon) \longrightarrow L_{n-1}(A,\epsilon) \longrightarrow \cdots.$$

Special cases of these sequences were obtained by Milnor and Husemoller, Karoubi, Pardon, Smith, Carlsson and Milgram. Let $G\pi : D(A) \to D(A)$ be the functor of Proposition 6.1 of [3], with D(A) the derived category of A. For any bounded f.g. projective A-module chain complex C the natural A-module chain map

$$\varinjlim_{(B,\beta)} B = G\pi(C) \longrightarrow \sigma^{-1}C$$

induces morphisms

$$\sigma^{-1}Q^* : \varinjlim_{(B,\beta)} Q^n(B,\epsilon) = Q^n(G\pi(C),\epsilon) \longrightarrow Q^n(\sigma^{-1}C,\epsilon),$$

$$\sigma^{-1}Q_* : \varinjlim_{(B,\beta)} Q_n(B,\epsilon) = Q_n(G\pi(C),\epsilon) \longrightarrow Q_n(\sigma^{-1}C,\epsilon)$$

with the direct limits taken over all the bounded f.g. projective A-module chain complexes B with a chain map $\beta : C \to B$ such that $\sigma^{-1}\beta : \sigma^{-1}C \to \sigma^{-1}B$ is a $\sigma^{-1}A$-module chain equivalence. The natural projection $D \otimes_A D \to D \otimes_{\sigma^{-1}A} D$ is an isomorphism for any bounded f.g. projective $\sigma^{-1}A$-module chain complex D (since this is already the case for D =

$\sigma^{-1}A$), so the Q-groups of $\sigma^{-1}C$ are the same whether $\sigma^{-1}C$ is regarded as an A-module or $\sigma^{-1}A$-module chain complex.

Theorem 2.4: (See Vogel [9, Theorem 8.4].) For any ring with involution A and noncommutative localization $\sigma^{-1}A$ the morphisms

$$\sigma^{-1}\Gamma_* : \Gamma_n(A \longrightarrow \sigma^{-1}A, \epsilon) \longrightarrow L_n^I(\sigma^{-1}A, \epsilon); \quad (C, \psi) \longmapsto \sigma^{-1}(C, \psi)$$

Proof: By algebraic surgery below the middle dimension it suffices to consider only the special cases n = 0, 1. In effect, it was proved in [9] that $\sigma^{-1}Q_*$ is an isomorphism for 0- and 1-dimensional C.

It was claimed in Proposition 25.4 of Ranicki [6] that $\sigma^{-1}\Gamma^*$ is also an isomorphism, assuming (incorrectly) that the chain complex lifting problem can always be solved. However, we do have:

Theorem 2.5: If $\sigma^{-1}A$ is a noncommutative localization of a ring with involution A which is stably flat over A, there is a localization exact sequence of ε-symmetric L-groups

$$\cdots \longrightarrow L^n(A, \epsilon) \longrightarrow L_I^n(\sigma^{-1}A, \epsilon) \xrightarrow{\partial} L^n(A, \sigma, \epsilon) \longrightarrow L^{n-1}(A, \epsilon) \longrightarrow \cdots.$$

Proof: For any bounded f.g. projective A-module chain complex C the natural A-module chain map $G\pi(C) \to \sigma^{-1}C$ induces isomorphisms in homology

$$H_*(G\pi(C)) \cong H_*(\sigma^{-1}C).$$

Thus the natural $\mathbb{Z}$ ($\mathbb{Z}_2$) module chain map

$$G\pi(C) \otimes_A G\pi(C) \longrightarrow \sigma^{-1}C \otimes_A \sigma^{-1}C = \sigma^{-1}C \otimes_{\sigma^{-1}A} \sigma^{-1}C$$

induces isomorphisms of ε-symmetric Q-groups

$$\sigma^{-1}Q^*: \varinjlim_{(B,\beta)} Q^n(B,\epsilon) \longrightarrow Q^n(\sigma^{-1}C,\epsilon)$$

(and also isomorphisms $\sigma^{-1}Q^*$ of ε-quadratic Q-groups). By Theorem 0.1 every n-dimensional induced f.g. projective $\sigma^{-1}A$-module chain complex D is chain equivalent to $\sigma^{-1}C$ for an ndimensional f.g. projective A-module chain complex C, with

$$Q^n(D,\epsilon) = Q^n(\sigma^{-1}C,\epsilon) = \varinjlim_{(B,\beta)} Q^n(B,\epsilon).$$

It follows that the morphisms of ε-symmetric Γ- and L-groups

$$\sigma^{-1}\Gamma^*: \Gamma^n(A \longrightarrow \sigma^{-1}A,\epsilon) \longrightarrow L_I^n(\sigma^{-1}A,\epsilon); \quad (C,\phi) \longmapsto \sigma^{-1}(C,\phi)$$

are also isomorphisms, and the localization exact sequence is given by Proposition 2.3.

Hypothesis 2.6: For the remainder of this section, we assume Hypothesis 2.1 and also that $A \rightarrow \sigma^{-1}A$ is an injection.

As in Proposition 2.2 of [2] it follows that all the morphisms in σ are injections. We shall now generalize the results of Ranicki [5] and Vogel [8] to prove that under Hypotheses 2.1, 2.6 the relative L-groups $L^*(A, \sigma, \varepsilon)$, $L_*(A, \sigma, \varepsilon)$ in the L-theory localization exact sequences are the L-groups of H (A, σ) with respect to the following duality involution. Define the torsion dual of an (A, σ)-module M to be the (A, σ)-module

$$M^{\wedge} = \mathrm{Ext}_A^1(M,A),$$

using the involution on A to define the left A-module structure. If M has f.g. projective A module resolution

$$0 \longrightarrow P_1 \xrightarrow{s} P_0 \longrightarrow M \longrightarrow 0$$

with s ∈ σ the torsion dual M^ has the dual f.g. projective A-module resolution

$$0 \longrightarrow P_0^* \xrightarrow{s^*} P_1^* \longrightarrow M\hat{} \longrightarrow 0$$

with s* ∈ σ.

Proposition 2.7: Let M = coker(s:P_1 → P_0), N = coker(t : Q_1 → Q_0) be (A, σ)-modules.

(i) The adjoint of the pairing

$$M \times M\hat{} \longrightarrow \sigma^{-1}A/A; \quad (g \in P_0, f \in P_1^*) \longmapsto fs^{-1}g$$

defines a natural A-module isomorphism

$$M\hat{} \longrightarrow \mathrm{Hom}_A(M, \sigma^{-1}A/A); \quad f \longmapsto (g \longmapsto fs^{-1}g).$$

(ii) The natural A-module morphism

$$M \longrightarrow M\hat{}\hat{}; \quad x \longmapsto (f \longmapsto \overline{f(x)})$$

is an isomorphism.

(iii) There are natural identifications

$$M \otimes_A N = \mathrm{Tor}_0^A(M, N) = \mathrm{Ext}_A^1(M\hat{}, N) = H_0(P \otimes_A Q),$$
$$\mathrm{Hom}_A(M\hat{}, N) = \mathrm{Tor}_1^A(M, N) = \mathrm{Ext}_A^0(M\hat{}, N) = H_1(P \otimes_A Q).$$

The functions

$$M \otimes_A N \longrightarrow N \otimes_A M; \quad x \otimes y \longmapsto y \otimes x,$$
$$\mathrm{Hom}_A(M\hat{}, N) \longrightarrow \mathrm{Hom}_A(N\hat{}, M); \quad f \longmapsto f\hat{}$$

determine transposition isomorphisms

$$T : \mathrm{Tor}_i^A(M, N) \longrightarrow \mathrm{Tor}_i^A(N, M) \quad (i = 0, 1).$$

(iv) For any finite subset $V = \{v_1, v_2, \ldots, v_k\} \subset M \otimes_A N$ there exists an exact sequence of (A, σ)-modules

$$0 \longrightarrow N \longrightarrow L \longrightarrow \bigoplus_k M^{\wedge} \longrightarrow 0$$

such that $V \subset \ker(M \otimes_A N \dashrightarrow M \otimes_A L)$.

***Proof*:**

(i) Apply the snake lemma to the morphism of short exact sequences

$$\begin{array}{ccccccccc} 0 & \longrightarrow & \mathrm{Hom}_A(P_0, A) & \longrightarrow & \mathrm{Hom}_A(P_0, \sigma^{-1}A) & \longrightarrow & \mathrm{Hom}_A(P_0, \sigma^{-1}A/A) & \longrightarrow & 0 \\ & & \downarrow s^* & & \downarrow s_1^* & & \downarrow s_2^* & & \\ 0 & \longrightarrow & \mathrm{Hom}_A(P_1, A) & \longrightarrow & \mathrm{Hom}_A(P_1, \sigma^{-1}A) & \longrightarrow & \mathrm{Hom}_A(P_1, \sigma^{-1}A/A) & \longrightarrow & 0 \end{array}$$

with s^* injective, s_1^* an isomorphism and s_2^* surjective, to verify that the A-module morphism

$$M^{\wedge} = \mathrm{coker}(s^*) \longrightarrow \mathrm{Hom}_A\left(M, \sigma^{-1}A/A\right) = \ker\left(s_2^*\right)$$

is an isomorphism.

(i) Immediate from the identification

$$s^{**} = s \,:\, (P_0)^{**} = P_0 \longrightarrow (P_1)^{**} = P_1 .$$

(ii) Exercise for the reader.

(iii) Lift each $v_i \in M \otimes_A N$ to an element

$$v_i \in P_0 \otimes_A Q_0 = \mathrm{Hom}_A\left(P_0^*, Q_0\right) \quad (1 \leqslant i \leqslant k).$$

The A-module morphism defined by

$$u = \begin{pmatrix} s^* & 0 & 0 & \ldots & 0 \\ 0 & s^* & 0 & \ldots & 0 \\ 0 & 0 & s^* & \ldots & 0 \\ \vdots & \vdots & \vdots & \ddots & \vdots \\ v_1 & v_2 & v_3 & \ldots & t \end{pmatrix} : U_1 = \left(\bigoplus_k P_0^*\right) \oplus Q_1 \longrightarrow U_0 = \left(\bigoplus_k P_1^*\right) \oplus Q_0$$

is in σ , so that L = coker(u) is an (A, σ)-module with an f.g. projective A-module resolution

$$0 \longrightarrow U_1 \xrightarrow{u} U_0 \longrightarrow L \longrightarrow 0.$$

The short exact sequence of 1-dimensional f.g. projective A-module chain complexes

$$0 \longrightarrow Q \longrightarrow U \longrightarrow \bigoplus_k P^{1-*} \longrightarrow 0$$

is a resolution of a short exact sequence of (A, σ)-modules

$$0 \longrightarrow N \longrightarrow L \longrightarrow \bigoplus_k M^\wedge \longrightarrow 0.$$

The first morphism in the exact sequence

$$\mathrm{Tor}_1^A\Big(M, \bigoplus_k M^\wedge\Big) \longrightarrow M \otimes_A N \longrightarrow M \otimes_A L \longrightarrow M \otimes_A \Big(\bigoplus_k M^\wedge\Big) \longrightarrow 0$$

sends $1_i \in \mathrm{Tor}_1^A(M, \oplus_k M^\wedge) = \oplus \mathrm{Hom}_A(M^\wedge, M^\wedge)$ to $v_i \in \ker(M \otimes_A N \to M \otimes_A L)$

Given an (A, σ)-module chain complex C define the ε-symmetric (resp. ε-quadratic) torsion Q-groups of C to be the $\mathbb{Z}_2$-hypercohomology (resp. $\mathbb{Z}_2$-hyperhomology) groups of the -transposition involution $T_\varepsilon = {}_\varepsilon T$ on the $\mathbb{Z}$-module chain complex $\mathrm{Tor}_1^A(C, C) = \mathrm{Hom}_A(C^\wedge, C)$

$$Q^n_{\mathrm{tor}}(C, \epsilon) = H^n\big(\mathbb{Z}_2; \mathrm{Tor}_1^A(C, C)\big) = H_n\big(\mathrm{Hom}_{\mathbb{Z}[\mathbb{Z}_2]}(W, \mathrm{Tor}_1^A(C, C))\big),$$

$$Q_n^{\mathrm{tor}}(C, \epsilon) = H_n\big(\mathbb{Z}_2; \mathrm{Tor}_1^A(C, C)\big) = H_n\big(W \otimes_{\mathbb{Z}[\mathbb{Z}_2]} \big(\mathrm{Tor}_1^A(C, C)\big)\big).$$

There are defined forgetful maps

$$1+T_\epsilon : Q_n^{tor}(C,\epsilon) \longrightarrow Q^n_{tor}(C,\epsilon); \quad \psi \longmapsto (1+T_\epsilon)\psi,$$

$$Q^n_{tor}(C,\epsilon) \longrightarrow H_n\left(\mathrm{Tor}_1^A(C,C)\right); \quad \phi \longmapsto \phi_0.$$

The element $\varphi_0 \in H_n(\mathrm{Tor}_1^A\ (C, C))$ is a chain homotopy class of A-module chain maps φ_0: $C^{n-\wedge} \rightarrow C$.

An n-dimensional ε-symmetric complex over (A, σ) (C, φ) is a bounded (A, σ)-module chain complex C together with an element $\varphi \in Q^n_{tor}$ (C, ε). The complex (C, φ) is Poincaré if the Amodule chain maps φ_0: $C^{n-\wedge} \rightarrow C$ are chain equivalences.

Example 2.8: A 0-dimensional ε-symmetric Poincaré complex (C, φ) over (A, σ) is essentially the same as a nonsingular ε-symmetric linking form (M, λ) over (A, σ), with M = $(C_0)^\wedge$ an (A, σ)-module and

$$\lambda = \phi_0 : M \times M \longrightarrow \sigma^{-1}A/A$$

a sesquilinear pairing such that the adjoint

$$M \longrightarrow M^\wedge; \quad x \longmapsto \left(y \longmapsto \lambda(x,y)\right)$$

is an A-module isomorphism.

The n-dimensional torsion ε-symmetric L-group L^n_{tor} (A, σ, ε) is the cobordism group of ndimensional ε-symmetric Poincaré complexes (C, φ) over (A, σ), with C n-dimensional. In particular, L^0_{tor} (A, σ, ε) is the Witt group of nonsingular ε-symmetric linking forms over (A, σ).

Similarly in the ε-quadratic case, with torsion L-groups L_n^{tor} (A, σ, ε). The ε-quadratic torsion L-groups are 4-periodic

$$L_n^{\text{tor}}(A,\sigma,\epsilon) = L_{n+2}^{\text{tor}}(A,\sigma,-\epsilon) = L_{n+4}^{\text{tor}}(A,\sigma,\epsilon).$$

Theorem 2.9: If $A \to \sigma^{-1}A$ is injective the relative L-groups in the localization exact sequences of Proposition 2.3

$$\cdots \longrightarrow L^n(A,\epsilon) \longrightarrow \Gamma^n(A \longrightarrow \sigma^{-1}A,\epsilon) \xrightarrow{\partial} L^n(A,\sigma,\epsilon) \longrightarrow L^{n-1}(A,\epsilon) \longrightarrow \cdots,$$

$$\cdots \longrightarrow L_n(A,\epsilon) \longrightarrow \Gamma_n(A \longrightarrow \sigma^{-1}A,\epsilon) \xrightarrow{\partial} L_n(A,\sigma,\epsilon) \longrightarrow L_{n-1}(A,\epsilon) \longrightarrow \cdots$$

are the torsion L-groups

$$L^*(A,\sigma,\epsilon) = L^*_{\text{tor}}(A,\sigma,\epsilon),$$
$$L_*(A,\sigma,\epsilon) = L_*^{\text{tor}}(A,\sigma,\epsilon).$$

Proof: For any bounded (A, σ)-module chain complex T there exists a bounded f.g. projective A-module chain complex C with a homology equivalence $C \to T$. Working as in [8] there is defined a distinguished triangle of $\mathbb{Z}[\mathbb{Z}_2]$-module chain complexes

$$\Sigma \operatorname{Tor}_1^A(T,T) \longrightarrow C \otimes_A C \longrightarrow T \otimes_A T \longrightarrow \Sigma^2 \operatorname{Tor}_1^A(T,T)$$

with $\mathbb{Z}_2$ acting by the ε-transposition T_ε on the $\mathbb{Z}$-module chain complex $\operatorname{Tor}_1^A(T,T)$ and by the $(-\varepsilon)$-transpositions $T_{-\varepsilon}$ on $C \otimes_A C$ and $T \otimes_A T$, inducing long exact sequences

$$\cdots \longrightarrow Q^n_{\text{tor}}(T,\epsilon) \longrightarrow Q^{n+1}(C,-\epsilon) \longrightarrow Q^{n+1}(T,-\epsilon) \longrightarrow Q^{n-1}_{\text{tor}}(T,\epsilon) \longrightarrow \cdots.$$

$$\cdots \longrightarrow Q_n^{\text{tor}}(T,\epsilon) \longrightarrow Q_{n+1}(C,-\epsilon) \longrightarrow Q_{n+1}(T,-\epsilon) \longrightarrow Q_{n-1}^{\text{tor}}(T,\epsilon) \longrightarrow \cdots.$$

Passing to the direct limits over all the bounded (A, σ)-module chain complexes U with a homology equivalence $\beta: T \to U$ use Proposition 2.7(iv) to obtain

$$\varinjlim_{(U,\beta)} Q^{n+1}(U,-\epsilon)=0,$$

$$\varinjlim_{(U,\beta)} Q_{n+1}(U,-\epsilon)=0$$

and hence

$$\varinjlim_{(U,\beta)} Q^{n}_{\text{tor}}(U,\epsilon)=Q^{n+1}(C,-\epsilon),$$

$$\varinjlim_{(U,\beta)} Q_{n}^{\text{tor}}(U,\epsilon)=Q_{n+1}(C,-\epsilon).$$

Remark 2.10: The identification $L_*(A, \sigma, \varepsilon) = L_*^{\text{tor}}(A, \sigma, \varepsilon)$ for noncommutative $\sigma^{-1}A$ was first obtained by Vogel [8].

REFERENCES

1. Amnon Neeman, Triangulated Categories, Ann. of Math. Stud., vol. 148, Princeton University Press, 2001.
2. Amnon Neeman, Noncommutative localisation in algebraic K-theory II, Adv. Math. 213 (2007) 785–819.
3. Amnon Neeman, Andrew Ranicki, Noncommutative localisation in algebraic K-theory I, Geom. Topol. 8 (2004) 1385–1425.
4. Amnon Neeman, Andrew Ranicki, Aidan Schofield, Representations of algebras as universal localizations, Math. Proc. Cambridge Philos. Soc. 136 (2004) 105–117.
5. Andrew Ranicki, Exact Sequences in the Algebraic Theory of Surgery, Math. Notes, vol. 26, Princeton, 1981.
6. Andrew Ranicki, High Dimensional Knot Theory, Springer Monogr. Math., Springer, 1998.
7. Andrew Ranicki, Noncommutative localization in topology, in: Proc. 2002 ICMS Conference on Noncommutative Localization in Algebra and Topology, in: London Math. Soc. Lecture Note Ser., vol. 330, Cambridge Univ. Press, 2006, pp. 81–102.

8. Pierre Vogel, Localization in algebraic L-theory, in: Proc. 1979 Siegen Topology Symposium, in: Lecture Notes in Math., vol. 788, Springer-Verlag, 1980, pp. 482–495.
9. Pierre Vogel, On the obstruction group in homology surgery, Publ. Math. Inst. Hautes Etudes Sci. 55 (1982) 165–206.

CITATION

1. Andrew Ranicki, Noncommutative localization in algebraic L-theory, Advances in Mathematics, Volume 220, Issue 3, 15 February 2009, Pages 894-912, ISSN 0001-8708, http://dx.doi.org/10.1016/j.aim.2008.10.003.

Intersection Multiplicity over Noncommutative Algebras

Izuru Mori
Department of Mathematics, Purdue University,
West Lafayette, IN 47907, USA

5

ABSTRACT

In this paper, we will prove Serre's multiplicity conjectures for large class of connected algebras (not necessarily commutative), using the intersection multiplicity defined by Ext groups instead of Tor groups. Since this new definition of the intersection multiplicity agrees with classical Serre's definition for nice commutative rings, the result is a noncommutative version of that of Peskine and Szpiro. The result also applies to noncommutative filtered algebras in certain situations.

INTERSECTION MULTIPLICITIES

Let R be a noetherian ring and M, N be finitely generated R-modules. For a commutative regular local ring R, Serre defined the intersection multiplicity of M and N by

$$\chi(M, N) := \sum_{i=0}^{\infty} (-1)^i \operatorname{length} \operatorname{Tor}_i^R(M, N)$$

When length $(M \otimes_R N) < \infty$. We use this definition for any commutative ring R, not necessarily a regular local ring, when it is well defined. Serre's multiplicity conjectures are:

Conjecture 1.1 (Serre, 1961 [1]): Let R be a noetherian commutative regular local ring and M, N be finitely generated R-modules such that length $(M \otimes_R N) < \infty$. Then

- (Dimension) Kdim M + Kdim N $\leq$ Kdim R
- (Vanishing) If Kdim M + Kdim N < Kdim R, then $\chi(M, N) = 0$.
- (Positivity) If Kdim M + Kdim N = Kdim R, then $\chi(M, N) > 0$.

Serre proved the dimension conjecture in general and the other conjectures for a noetherian commutative regular local ring containing a field (1961). The vanishing conjecture was proved by Roberts [2] and Gillet and Soulé [3] independently (1985). In fact, they proved it for a noetherian commutative local complete intersection ring R and finitely generated R-modules M, N of finite projective dimension. The positivity conjecture is still open. If R, M, N are all graded, then a stronger version of Serre's multiplicity conjectures holds by Peskine and Szpiro [4] (1974).

Now we would like to extend the definition of the intersection multiplicity to noncommutative rings. Since Tor groups of two right modules do not make sense but Ext groups do, we simply replace all Tor groups by Ext groups in the definition. Motivated by [5], our new candidate for the intersection multiplicity for a commutative ring R is

$$\xi(M,N) := \sum_{i=0}^{\infty}(-1)^i \text{ lenght } \operatorname{Ext}^i_R(M,N).$$

In fact, if R is a nice commutative ring, then ξ agrees with χ up to sign by Chan.

Theorem 1.2 [6, Theorem 4]: Let R be a noetherian commutative local complete intersection ring, and M, N be finitely generated R-modules of finite projective dimension. If length $(M \otimes_R N) < \infty$ and Kdim M + Kdim N $\leq$ Kdim R, then

$$\chi(M,N) = (-1)^{\operatorname{Kdim} N}\xi(M,N)$$

By the above theorem, we would like to adopt ξ as a definition of the intersection multiplicity for noncommutative rings (up to sign), but here is another obstacle: if R is not commutative and M, N are right R-modules, then Ext^i_R R(M, N) is no longer an R-module so length Ext^i_R (M, N) does not make sense. However, if R is an algebra over a field k, then Ext^i_R (M, N) has a k-vector space structure so dim k Ext^i_R (M, N) makes sense. In fact, if R is a commutative local ring containing the residue field k, then length and k-vector space dimension of an R-module agree. Moreover, if R is a commutative finitely generated kalgebra and k is algebraically closed, then again length and k-vector space dimension of an R-module agree by Hilbert Nullstellensatz. So we define the intersection multiplicity for noncommutative k-algebras (up to sign) as follows:

Definition 1.3: Let R be a noetherian k-algebra and M be a finitely generated right R-module

(1) For a finitely generated left R-module N, we define

$$\chi(M,N)=\sum_{i=0}^{\infty}(-1)^i \dim_k \mathrm{Tor}_i^R(M,N)$$

(2) For a finitely generated right R-module N, we define

$$\xi(M,N)=\sum_{i=0}^{\infty}(-1)^i \dim_k \mathrm{Ext}^i_R(M,N)$$

Of course, χ and ξ are not always well defined. For example, ξ (M, N) is well defined if and only if

(1) Ext^i_R (M, N) = 0 for all i ≫ 0, and

dimk Ext^i_R (M, N) < ∞ for all i ≥ 0.

We define

$$r(M,N)=\sup\{i \mid \mathrm{Ext}^i_R(M,N)\neq 0\}$$

so that the first condition is equivalent to r(M, N) < ∞. Note that r (M, N) ≤ sup {projdim M, injdim N} but the inequality could be strict. If R has finite global dimension, then r (M, N) < ∞ for all M, N.

In this paper, we will prove Serre's multiplicity conjectures for connected algebras using these definitions of the intersection multiplicity.

SOME PROPERTIES OF CONNECTED ALGEBRAS

Throughout let k be a fixed field. A connected k-algebra is a graded algebra of the form $A = A_0 \oplus A_1 \otimes A_2 \otimes \cdots$ where $A_0 = k$. It is known that there are strong analogies between connected algebras and commutative local rings. In this section, we will define some properties of connected algebras. First, we will extend some homological properties of commutative local rings to connected algebras.

Let A be a connected algebra. One reasonable assumption on A to have good intersection theory is that the intersection multiplicity of "the origin" and "the whole space" is 1. It is clear that $\chi(A, k) = \chi(k, A) = \xi(A, k) = 1$. We expect that $\xi(k, A)$ is also 1 up to sign, but $\xi(k, A)$ is not always well defined for a general connected algebra A. We will impose some conditions on A so that $\xi(k, A)$ is well defined. Recall that there are two conditions in order for $\xi(k, A)$ to be well defined. The first condition $r(k, A) < \infty$ is guaranteed if A has finite injective dimension as a graded right module over itself. The second condition dimk $\mathrm{Ext}^i_A(k, A) < \infty$ for all $i \geq 0$ is guaranteed by the condition χ defined below:

Definition 2.1: Let A be a noetherian connected k-algebra. We say that

A satisfies χ on the right if $\dim_k \mathrm{Ext}^i_A(k, M) < \infty$ for all $I \geq 0$ and for all finitely generated graded right A-modules M. The condition χ on the left is similarly defined.

The condition χ is rather mild. Every noetherian commutative connected algebra satisfies it, and so do most noetherian noncommutative connected algebras of importance. The condition χ is essential to get a theory for noncommutative algebraic geometry which resembles the commutative theory. See [7] for details.

Definition 2.2: A noetherian connected algebra A is called AS-Gorenstein on the right if A satisfies χ on both sides and has finite injective dimension as a graded right module over itself.

A noetherian connected algebra A is called AS-regular if A satisfies χ on both sides and has finite global dimension.

By definition, every AS-regular algebra is AS-Gorenstein on both sides. If A is AS-Gorenstein, then $\xi(k, A)$ is well defined. Under mild assumptions, the converse is also true ([8, Theorem 4.5], [9, Theorem 1.2]). This suggests that AS-Gorenstein algebras are most reasonable algebras to have good intersection theory.

Next, we will define an AS Cohen–Macaulay algebra.

Definition 2.3: Let A be a connected algebra and M be a graded right A-module. The i-th local cohomology group of M is

$$H_{\mathfrak{m}}^{i}(M) = \lim_{n\to\infty} \mathrm{Ext}_{A}^{i}(A/A_{\geq n}, M)$$

We define the local dimension of M by

$$\mathrm{ldim}\, M = \sup\{i \mid H_{\mathfrak{m}}^{i}(M) \neq 0\}$$

Suppose that ldim A = d < ∞. A right dualizing module is a graded A-A bimodule w_A such that, for each i, there is a functorial isomorphism

$$\mathrm{Ext}_{A}^{i}(M, w_A) \cong H_{\mathfrak{m}}^{d-i}(M)^{*}$$

of graded left A-modules for all finitely generated graded right A-modules M, where $H_{m}^{d-i}(m)^{*}$ is the Matlis dual of $H_{m}^{d-i}(m)$.

A noetherian connected algebra is called AS Cohen–Macaulay on the right if A has a right dualizing module w_A.

Since $A/A_{\geq n}$ is a graded A-A bimodule, we can also define these notions for a graded left A-module M.

By definition, it is clear that $w_A \cong H^d_m(A)^*$ as graded left A-modules.

Lemma 2.4: Let A be a connected algebra with ldim A = d < ∞. If A has a right dualizing module wA, then the injective dimension of w_A as a graded right A-module is d.

Proof: By [10, Proposition 1.7], the injective dimension of w_A can be calculated by

$$\sup\{i \mid \operatorname{Ext}^i_A(M, w_A) \neq 0 \text{ for all finitely generated graded right } A\text{-modules } M\}.$$

But for every finitely generated graded right A-module M,

$$\operatorname{Ext}^i_A(M, w_A) \cong H^{d-i}_{\mathfrak{m}}(M)^* = 0$$

if i>d, and $\operatorname{Ext}^d_A(k, w_A) \cong H^0_m(k)^* \cong k \neq 0$, hence the result.

It is known that if A is AS-Gorenstein algebra, then A is AS Cohen–Macaulay, and a dualizing module w_A is isomorphic to a shift of A as graded left and right A-modules.

Finally, we will define some numerical properties of connected algebras.

Definition 2.5: Let V be a graded k-vector space. We define a(V) = sup{i | $V_i \neq 0$} and b(V) = inf{i | $V_i \neq 0$}. We say that V is bounded above (respectively below) if a (V) < ∞ (respectively b (V) > ∞). We say that V is locally finite if $\dim_k V_i < \infty$ for all i. For a locally finite graded k-vector space, we define the Hilbert series of V by

$$H_V(t) = \sum_{i=-\infty}^{\infty} (\dim_k V_i) t^i \in \mathbf{Z}[[t, t^{-1}]],$$

And GKdimension of V by

GKdimV = inf{$\rho \in R \mid \dim_k V_i \leq i^{\rho-1}$ for all $i \gg 0$}.

The following lemma is useful in this paper.

Lemma 2.6: Let A be a right noetherian connected algebra, and M be a finitely generated graded right A-module. Then

(1) M is locally finite and bounded below.
(2) Let N be a graded left A-module. If N is locally finite, then so is Tor_i^A (M, N) for all i$\geq$ 0. If N is bounded below, then so is Tor_i^A (M, N) for all i$\geq$ 0. In fact, b(Tor_i^A (M, N)) $\geq$ b(M) + b(N) + i for all i $\geq$0.
(3) Let N be a graded right A-module. If N is locally finite, then so is Ext_A^i (M, N) for all i$\geq$ 0. If N is bounded below, then so is Ext_A^i (M, N) for all i $\geq$0.
(4) Let S be a left noetherian connected algebra, and N be a graded S-A bimodule. If N is finitely generated as a graded left S-module, then Ext_A^i (M, N) is a finitely generated graded left S-module for all i $\geq$0.

Proof: (1) follows from [7, Proposition 2.1]. (2) and (3) follows from (1) by considering a minimal finitely generated free resolution of M. (4) follows from [7, Proposition 3.1 (4)].

We are mainly interested in a graded module M such that GKdimM $< \infty$ and $H_M(t)$ is a rational function over the complex number. In this case, GKdimM is the order of the pole of $H_M(t)$ at t = 1, which is an integer. If A is a noetherian commutative connected algebra and M is a finitely generated graded A-module, then $H_M(t)$ is a rational function over the complex number and GKdimM = KdimM $< \infty$. For such a graded module, we will define the multiplicity.

Definition 2.7: Let V be a locally finite graded k-vector space such that GKdimV $< \infty$ and H_V (t) is a rational function over the complex number. The multiplicity series of V is defined by

$$e_V(t) = (1-t)^{\mathrm{GKdim}\,V} H_V(t),$$

And the multiplicity of V is defined by

$$e(V) = \lim_{t \to 1} e_V(t).$$

The following definition is inspired by [11, Definition 1.4].

Definition 2.8: Let A be a noetherian connected algebra. A graded right A-module M is called Γ -finite if

- ldimM < ∞, and
- H_m^i (M) is locally finite and bounded above for all i ≥ 0.

We say that A is universally rational on the right if, for every finitely generated graded right A-module M, M is Γ -finite, and both $H_M(t)$ and

$$\sum_{i=0}^{\infty}(-1)^i H_{H_m^i(M)}(t)$$

Are rational functions over the complex number and they are equal as rational functions.

Note that Γ -finiteness condition allows us to define $\sum_{i=0}^{\infty}(-1)^i H_{H_m^i(M)}(t)$ as an element of $Z[[t-1]][t]$. We do not know if every finitely generated graded right A-module is Γ -finite for a general right noetherian connected algebra A. If A is AS Cohen–Macaulay on the right and w_A is a finitely generated graded A-module on either side, then every finitely generated graded A-module M is Γ - finite because $H_m^i(M) \cong \mathrm{Ext}_A^{d-i}(M, w_A)^*$ is locally finite and bounded above for all i ≥ 0 by Lemma 2.6 (3) and zero for i>d.

Although the definition is rather technical, many noetherian connected algebras of importance are universally rational. For example, the following algebras are universally rational by [11, Proposition 5.5]:

- an AS-regular algebra,
- an FBN AS-Gorenstein algebra,
- a noetherian connected PI algebra, and
- a graded quotient algebra of any of the above algebras.

In particular, a noetherian commutative connected algebra is universally rational.

INTERSECTION MULTIPLICITY OVER CONNECTED ALGEBRAS

In this section, we will prove Serre's multiplicity conjectures for connected algebras.

Definition 3.1: Let A be a noetherian connected k-algebra and M be a finitely generated graded right A-module.

(1) For a locally finite, bounded below, graded left A-module N, we define

$$\chi_{M,N}(t) = \sum_{i=0}^{\infty}(-1)^i H_{\mathrm{Tor}_i^A(M,N)}(t).$$

(2) For a finitely generated graded right A-module N such that r(M, N) < ∞, we define

$$\xi_{M,N}(t) = \sum_{i=0}^{\infty}(-1)^i H_{\mathrm{Ext}_A^i(M,N)}(t).$$

The definition of $\chi_{M,N}(t)$ involves an infinite sum of power series. However, in the setting of the definition, $\mathrm{Tor}_i^A(M, N)$ is locally finite for all i ≥ 0, and, for each n ∈ Z, $\mathrm{Tor}_i^A(M, N)_n = 0$ for all i ≫ 0 by Lemma 2.6 (2), so $\chi_{M,N}(t)$ is well defined.

On the other hand, in the setting of the definition, $\mathrm{Ext}_A^i(M, N)$ is locally finite for all i ≥ 0 by Lemma 2.6 (3), but we do not know if $\xi_{M,N}(t)$ is well defined in general without the condition r(M, N) < ∞.

It is clear that

$$\chi(M, N) = \lim_{t\to 1}\chi_{M,N}(t),$$

And

$$\xi(M, N) = \lim_{t\to 1}\xi_{M,N}(t)$$

When left hand side of each formula is well defined.

Lemma 3.2: Let A be a noetherian connected algebra. If M is a finitely generated graded right A-module, and N is a locally finite, bounded below, graded left Amodule, then

$$\chi_{M,N}(t) = H_M(t)H_N(t)/H_A(t).$$

Proof: Exactly the same proof as that of [12, Lemma 7] goes through.

By Lemma 3.2, Serre's multiplicity conjectures for χ (M, N) hold in the following sense.

Theorem 3.3: Let A be a noetherian connected k-algebra with GK-dimA < ∞, M be a finitely generated graded right A-module, and N be a finitely generated graded left A-module. Suppose that $H_A(t)$, $H_M(t)$, $H_N(t)$ are all rational functions, and χ (M, N) is well defined. Then

- (Dimension) GKdimM + GKdimN ≤ GKdimA
- (Vanishing) if GKdimM + GKdimN < GKdimA, then χ (M, N) = 0,
- (Positivity) if GKdimM + GKdimN = GKdimA, then

$$\chi(M, N) = e(M)e(N)/e(A) > 0.$$

Proof: This easily follows from Lemma 3.2, but see the proof of Theorem 3.9 below.

If A is a noetherian commutative connected algebra, then Theorem 3.3 improves the result of Peskine and Szpiro [4] in the sense that we do not need to assume that M or N has finite projective dimension in the statement. Moreover, Peskine and Szpiro used the Hilbert polynomial in their proof, so they needed an extra assumption, such as that A is generated by elements of degree 1 over A_0. If A is not commutative, then the Hilbert polynomial may not exist even if A is an AS-regular algebra generated by elements of degree 1 over k, so their proof has limited applications in noncommutative settings.

Now we focus on showing analogous results for ξ(M, N).

Lemma 3.4: Let S be a universally rational, AS Cohen–Macaulay algebra on the left with ldim_S S = d and S_w be a left dualizing module. If N is a finitely generated graded left S-module, then

$$\xi_{N,sw}(t) = (-1)^d H_N\left(t^{-1}\right).$$

Proof: Since S is universally rational and AS Cohen–Macaulay on the left, it follows that

$$\begin{aligned}\xi_{N,sw}(t) &:= \sum_{i=0}^{d}(-1)^d H_{\mathrm{Ext}_S^i(N,sw)}(t) = \sum_{i=0}^{d}(-1)^i H_{H_{\mathfrak{m}}^{d-i}(N)^*}(t) \\ &= \sum_{i=0}^{d}(-1)^i H_{H_{\mathfrak{m}}^{d-i}(N)}\left(t^{-1}\right) = (-1)^d \sum_{i=0}^{d}(-1)^i H_{H_{\mathfrak{m}}^{i}(N)}\left(t^{-1}\right) \\ &= (-1)^d H_N\left(t^{-1}\right). \qquad \square\end{aligned}$$

Definition 3.5: Let A be a noetherian connected algebra. We say that A has the property (P) on the right if there is a universally rational AS Cohen–Macaulay algebra S on the left, and a graded algebra homomorphism S → A such that A is finitely generated as a graded left S-module.

If A is a graded quotient algebra of an AS-regular algebra, or an FBN ASGorenstein algebra on the left, then A has the property (P) on the right. In particular, every noetherian commutative connected algebra has the property (P).

Proposition 3.6: Let A be a noetherian connected algebra having the property (P) on the right, and M be a finitely generated graded right A-module. If N is a graded A-A bimodule, finitely generated on the left, such that r(M, N) < ∞, then

$$\xi_{M,N}(t) = H_M\left(t^{-1}\right) H_N(t) / H_A\left(t^{-1}\right)$$

Proof: We will carefully follow the proof of [13, Theorem 1]. Let S be a universally rational AS Cohen–Macaulay algebra in the property (P) with $\mathrm{ldim}_S\, S = d$, and S_w be a left dualizing module for S. By Lemma 2.4, S_w has a finite minimal graded injective resolution I • as a graded left S-module. Let F• be a minimal graded free resolution of the graded right A-module M. Then there is a canonical isomorphism of complexes of graded k-vector spaces

$$F^{\bullet} \otimes_A \mathrm{Hom}_S(N, I^{\bullet}) \cong \mathrm{Hom}_S(\mathrm{Hom}_A(F^{\bullet}, N), I^{\bullet})$$

by [14, Lemma 4.4]. Filtering the complex on the left by the cohomological degree of F•, and the one on the right by that of I •, we obtain two spectral sequences

$$_2E_p^q = \mathrm{Tor}_p^A\left(M, \mathrm{Ext}_S^q(N,_S w)\right)$$

And

$$^2E^{p,q} = \mathrm{Ext}_S^q\left(\mathrm{Ext}_A^P(M, N),_S w\right)$$

Converging to a common limit.

Since S is universally rational, $S^w H_m^d(S)^*$ (isomorphic as graded right Smodules) is locally finite and bounded below. Since A is finitely generated as a graded left S-module, it follows that N is also finitely generated as a graded left S-module, so $\mathrm{Ext}_S^q(N,_S w)$ is locally finite and bounded below for all $q \geq 0$ by Lemma 2.6 (3). By the first spectral sequence, we have

$$\sum_{p,q}(-1)^{p+q} H_{\mathrm{Tor}_p^A(M,\mathrm{Ext}_S^q(N,_S w))}(t)$$

$$= \sum_q (-1)^q \sum_p (-1)^p H_{\mathrm{Tor}_p^A(M,\mathrm{Ext}_S^q(N,_S w))}(t)$$

$$= \sum_q (-1)^q \chi_{M,\mathrm{Ext}_S^q(N,_S w)}(t)$$

$$= \sum_q (-1)^q H_{\mathrm{Ext}_S^q(N,_S w)}(t) H_M(t)/H_A(t) \quad \text{[by Lemma 3.2]}$$

$$= \xi_{N,sw}(t) H_M(t)/H_A(t)$$

$$= (-1)^d H_N(t^{-1}) H_M(t)/H_A(t) \quad \text{[by Lemma 3.4]}.$$

On the other hand, since $\mathrm{Ext}_A^p(M,N)$ is a finitely generated graded left S-module by Lemma 2.6 (4), the second spectral sequence gives

$$\sum_{p,q} (-1)^{p+q} H_{\mathrm{Ext}_S^q(\mathrm{Ext}_A^p(M,N),sw)}(t)$$

$$= \sum_p (-1)^p \sum_q (-1)^q H_{\mathrm{Ext}_S^q(\mathrm{Ext}_A^p(M,N),sw)}(t)$$

$$= \sum_p (-1)^p \xi_{\mathrm{Ext}_A^p(M,N),sw}(t)$$

$$= (-1)^d \sum_p (-1)^p H_{\mathrm{Ext}_A^p(M,N)}(t^{-1}) \quad \text{[by Lemma 3.4]}$$

$$= (-1)^d \xi_{M,N}(t^{-1}).$$

Hence

$$\xi_{M,N}(t^{-1}) = H_N(t^{-1}) H_M(t)/H_A(t).$$

Theorem 3.7: Let A be a noetherian connected algebra, M, N be finitely generated graded right A-modules such that $r(M, N) < \infty$. If either

(1) M has finite projective dimension, or

(2) A has the property (P) on the right, $r(M, A) < \infty$, and N has finite flat dimension,

Then

$$\xi_{M,N}(t) = H_N(t^{-1})H_M(t)/H_A(t^{-1})$$

Note that if A is AS-Gorenstein, then r(M, A) < ∞ for all M. If M has finite projective dimension, then r(M, N) < ∞ for all N.

Proof: If M has finite projective dimension, then the result is classical. Suppose that A has the property (P), r(M, A) < ∞, and N has finite flat dimension. Applying Proposition 3.6 to N = A, we have

$$\xi_{M,A}(t) = H_M(t^{-1})H_A(t)/H_A(t^{-1})$$

Since N has finite flat dimension, by [15, Theorem 1.10], we have a convergent spectral sequence

$$_2E_p^q = \mathrm{Tor}_{-p}^A(N, \mathrm{Ext}_A^q(M, A)) \Rightarrow \mathrm{Ext}_A^{q+p}(M, N)$$

Of graded k-vector spaces. Since $\mathrm{Ext}_A^i(M, A)$ are locally finite, bounded below graded left A-modules for all i ≥ 0 by Lemma 2.6 (3), we have

$$\xi_{M,N}(t) := \sum_{i=0}^{\infty}(-1)^i H_{\mathrm{Ext}_A^i(M,N)}(t)$$

$$= \sum_{p,q}(-1)^{p+q} H_{\mathrm{Tor}_{-p}^A(N,\mathrm{Ext}_A^q(M,A))}(t)$$

$$= \sum_q (-1)^q \sum_p (-1)^p H_{\mathrm{Tor}_{-p}^A(N,\mathrm{Ext}_A^q(M,A))}(t)$$

$$= \sum_q (-1)^q \chi_{N,\mathrm{Ext}_A^q(M,A)}(t)$$

$$= \sum_q (-1)^q H_{\mathrm{Ext}_A^q(M,A)}(t) H_N(t)/H_A(t) \quad \text{[by Lemma 3.2]}$$

$$= \xi_{M,A}(t)H_N(t)/H_A(t)$$

$$= \left[H_M\left(t^{-1}\right)H_A(t)/H_A\left(t^{-1}\right)\right]\cdot H_N(t)/H_A(t)$$

$$= H_M\left(t^{-1}\right)H_N(t)/H_A\left(t^{-1}\right).$$

Remark 3.8: In the above proof,

$$\chi_{N,\mathrm{Ext}_A^q(M,A)}(t) = \sum_p (-1)^p H_{\mathrm{Tor}_{-p}^A(N,\mathrm{Ext}_A^q(M,A))}(t)$$

is well defined for each q ≥ 0 by Lemma 2.6 (2). This means that, fixing degree n and r$\geq$ 2, $\left({}_rE_p^q\right)_n = 0$ for all but finitely many pairs (p, q), so we can freely change the order of the sum

$$\sum_{p,q} (-1)^{p+q} H_{\mathrm{Tor}_{-p}^A(N,\mathrm{Ext}_A^q(M,A))}(t)$$

Similar remark applies to the proof of Proposition 3.6 above.

The following theorem is a version of Serre's conjectures for ξ(M, N).

Theorem 3.9: Let A be a noetherian connected k-algebra having the property (P) with GKdimA < ∞, and M, N be finitely generated graded right A-modules such that ξ(M, N) is well defined. Suppose that one of the followings is true:

(1) M has finite projective dimension,

(2) N has a graded A-A bimodule structure, finitely generated on both sides, or

(3) r(M, A) < ∞ and N has finite flat dimension.

Then

- (Dimension) GKdimM + GKdimN ≤ GKdimA,
- (Vanishing) if GKdimM + GKdimN < GKdimA, then ξ(M, N) = 0,
- (Positivity) if GKdimM + GKdimN = GKdimA, then

$$\xi(M, N) = (-1)^{\mathrm{GKdim}\, N} e(M)e(N)/e(A)$$

Proof: Write d = GKdimA, m = GKdimM, n = GKdimN. By Proposition 3.6 and Theorem 3.7,

$$\begin{aligned}
\xi(M, N) &= \lim_{t\to 1} \xi_{M,N}(t) \\
&= \lim_{t\to 1} \frac{H_M(t^{-1})H_N(t)}{H_A(t^{-1})} \\
&= \lim_{t\to 1} \frac{(1-t^{-1})^{-m} e_M(t^{-1})(1-t)^{-n} e_N(t)}{(1-t^{-1})^{-d} e_A(t^{-1})} \\
&= \lim_{t\to 1} \frac{[(-t^{-1})(1-t)]^{-m} e_M(t^{-1})(1-t)^{-n} e_N(t)}{[(-t^{-1})(1-t)]^{-d} e_A(t^{-1})} \\
&= \frac{e(M)e(N)}{e(A)} \lim_{t\to 1} \left(-t^{-1}\right)^{d-m} (1-t)^{d-m-n}
\end{aligned}$$

$$= (-1)^{d-m} \frac{e(M)e(N)}{e(A)} \lim_{t\to 1} (1-t)^{d-m-n},$$

Hence the result.

Definition 3.10: Let R be a right noetherian k-algebra and M, N be finitely generated right R-modules. We define the intersection multiplicity of M and N by M N = $(-1)^{\mathrm{GKdimN}}$ ξ(M, N), provided GKdimN is an integer. (See [16] for the definition of GKdimension for an ungraded module.)

Theorem 3.9 together with Theorem 3.3 shows that two definitions of the intersection multiplicity agree if A is commutative, which is analogous to Theorem 1.2.

Corollary 3.11. Let A be a noetherian commutative connected k-algebra and M, N be finitely generated graded A-modules. If both χ (M, N) and M N are well defined, then

$$\chi(M, N) = M \cdot N.$$

Note that we do not need to assume that M or N has finite projective dimension in Corollary 3.11.

INTERSECTION MULTIPLICITY OVER FILTERED ALGEBRAS

In this last section, we will see that Serre's multiplicity conjectures hold for noncommutative filtered algebras in certain situations. Main references to filtered algebras are [17] and [16]. We will treat the following type of filtrations:

Definition 4.1: Let C be a category and V be an object in C. A filtration of V is a family of subobjects of V , F V = {FpV: p ∈ Z}, such that

$$\cdots \subset F_{p-1}V \subset F_pV \subset F_{p+1}V \subset \cdots,$$

And $V = \bigcup_{p \in Z} F_pV$. We say that F V is discrete if there is an integer s such that $F_sV = 0$.

We will give examples of filtered rings in order to fix some terminologies.

Example 4.2: (1) Let R be a ring and I be a two-sided ideal of R. The I -adic filtration of R is obtained by putting $F_pR = R$ for $p \geq 0$ and $F_pR = I^{-p}$ for $p < 0$. If M is a right R-module, then the I -adic filtration of M is obtained by putting $F_pM = M$ for $p \geq 0$ and $F_pM = MI^{-p}$ for $p < 0$.

(2) Let R be a finitely generated k-algebra and V be a finite dimensional k-vector space generating R as a k-algebra. The standard filtration of R with respect to V is obtained by putting $F_pR = 0$ for $p < 0$ and $F_pR =$

$\sum_{i=0}^{p} V^i$ (by convention $V^0 = k$) for $p \geq 0$. If F R is a standard filtration of R, then it is clear that F R is discrete, and gr R is a connected k-algebra.

(3) Let R be a filtered ring, M be a finitely generated right R-module, and W be a finitely generated F_0R-submodule of M which generates M as a right R-module. The standard filtration of M with respect to W is obtained by putting $F_pM = WF_pR$ for all p. Suppose that FM is a standard filtration of M. It is clear that if F R is discrete, then so is FM. Moreover, if M is a finitely generated right R-module, then grM is a finitely generated graded right grR-module.

Let R be a filtered ring and M, N be filtered right R-modules. An R-module homomorphism $f \in \mathrm{Hom}_R(M, N)$ is said to have degree p if f $(F_qM) \subset F_{p+q}N$ for all $q \in Z$. Homomorphisms of degree p form a subgroup $F_p\ \mathrm{Hom}_R(M, N)$ of $\mathrm{Hom}_R(M, N)$. If M is finitely generated over R, then F HomR(M, N) is a filtration of HomR(M, N) in the category of abelian groups. There is a natural map

$$\varphi : \mathrm{gr}\,\mathrm{Hom}_R(M, N) \rightarrow \mathrm{Hom}_{\mathrm{gr}\,R}(\mathrm{gr}\,M, \mathrm{gr}\,N)$$

Given by $\varphi(f_p)(m_q) = f(m)_{p+q}$ for $f \in F_p\ \mathrm{Hom}_R(M, N)$ and $m \in F_qM$.

The existence of the spectral sequence in the proof of the following theorem is known (see [18]). We will recall the construction in order to check the convergence.

Theorem 4.3: Let R be a right noetherian filtered k-algebra such that grR is right noetherian, and M, N be finitely generated right R-modules such that grM, grN are finitely generated over grR. Suppose that F N is discrete. If ξ_{gr} R(grM, grN) is well defined, then $\xi_R(M, N)$ is also well defined and

$$\xi_R(M, N) = \xi_{\mathrm{gr}\,R}(\mathrm{gr}\,M, \mathrm{gr}\,N).$$

In particular, if we take standard filtrations on R, M, N (such that grR is right noetherian), then

$M \cdot N = \operatorname{gr} M, \operatorname{gr} N.$

Proof: We will follow [18] for the construction of the spectral sequence below. By [16, Theorem 7.6.17], there is a resolution $P\bullet \to M \to 0$ by finitely generated filtered-free modules such that $\operatorname{gr}P\bullet \to \operatorname{gr}M \to 0$ is a free resolution as graded grR-modules. Let $C\bullet$ be the complex $\operatorname{Hom}_R(P\bullet, N)$. Since differentials in the resolution $P\bullet \to M \to 0$ have degree 0, differentials in $C\bullet$ preserve degrees, so $C\bullet$ has a filtration $F\,C\bullet$ in the category of cochain complexes. By [19, Construction Theorem 5.4.1], $F\,C\bullet$ naturally determines a spectral sequence $_1E^{pq} = H^{p+q}(F_pC\bullet/F_{p-1}C\bullet)$. By [17, Lemma 6.4],

$$\begin{aligned} F_pC^n/F_{p-1}C^n &= F_p \operatorname{Hom}_R(P_n, N)/F_{p-1}\operatorname{Hom}_R(P_n, N) \\ &= (\operatorname{gr}\operatorname{Hom}_R(P_n, N))_p \\ &\cong \operatorname{Hom}_{\operatorname{gr} R}(\operatorname{gr} P_n, \operatorname{gr} N)_p, \end{aligned}$$

So $_1E^{pq} = \operatorname{Ext}^{p+q}_{\operatorname{gr}R}(\operatorname{gr}M, \operatorname{gr}N)_p$. If $F\,N$ is discrete, then so is $F\operatorname{Hom}_R(P_n, N)$ for each n by [17, Proposition 6.3], hence $_1E^{pq}$ converges to $H^{p+q}(C\bullet) = \operatorname{Ext}^{p+q}_R(M, N)$ by [19, Classical Convergence Theorem 5.5.1].

Since the spectral sequence converges as k-vector spaces, we have

$$\begin{aligned} \xi_{\operatorname{gr} R}(\operatorname{gr} M, \operatorname{gr} N) &= \sum_i (-1)^i \dim_k \operatorname{Ext}^i_{\operatorname{gr} R}(\operatorname{gr} M, \operatorname{gr} N) \\ &= \sum_i (-1)^i \sum_p \dim_k \operatorname{Ext}^i_{\operatorname{gr} R}(\operatorname{gr} M, \operatorname{gr} N)_p \\ &= \sum_i (-1)^i \sum_p \dim_k ({}_1E^{p,i-p}) \\ &= \sum_i (-1)^i \sum_p \dim_k ({}_\infty E^{p,i-p}) \\ &= \sum_i (-1)^i \dim_k \operatorname{Ext}^i_R(M, N) \\ &= \xi_R(M, N). \end{aligned}$$

Note that if $\xi_{grR}(grM, grN) = \sum_i(-1)^i \sum_p \dim_k({}_1E^{p,i-p})$ is well defined, then ${}_1E$ is bounded and ${}_1E_{pq}$ are all finite dimensional. Since ${}_\infty E^{pq}$ are subquotients of ${}_1E^{pq}$, it follows that $\sum_i(-1)i \sum_p \dim k({}_\infty E^{p,i-p})$ is also well defined. In particular, if we take standard filtrations on R, M, N (such that gr R is right noetherian), then GKdimN = GKdim grN by [16, Proposition 8.1.14], hence the result.

Example 4.4: Let R be the Weyl algebra, and V be the standard k-vector space generating R as a k-algebra. By the standard filtration on R with respect to V , A = grR is the polynomial ring. Let M and N be finitely generated R-modules. By taking standard filtrations on M, N, we have M N = grM grN when grM · grN is well defined by Theorem 4.3, and Serre's multiplicity conjectures hold for such M, N by Theorem 3.9.

Let R be the first Weyl algebra. For f ∈ R, we define the degree of f to be the integer p such that $f \in F_pR$ but $f \notin F_{p-1}R$, and the initial form $\bar{f}$ of f to be the image of f in $F_pR/F_{p-1}R$. If M = R/f R, N = $R/_gR$ for some f, g ∈ R, then we have grM = $A/(\bar{f})$, grN = $A/(\bar{g})$. It is easy to see that grM · grN is well defined if and only if $\gcd(\bar{f}, \bar{g}) = 1$, and in this case, $M \cdot N = grM \cdot grN = \deg\bar{f} \deg\bar{g} = \deg f \deg g$.

Finally, we will see that the spectral sequence in Theorem 4.3 may not be convergent if F N is not discrete as the following example shows:

Example 4.5: Let R = k[x,y] be the commutative polynomial ring. There are two natural filtrations of R, namely, I = (x, y)-adic filtration and the standard filtration with respect to V = kx + ky. In either filtration, grR = k[x,y] =: A is the polynomial ring. However, the standard filtration is better than the I - adic filtration in terms of calculating the intersection multiplicity. For example, let M = R/(x), and N = $R/(y + y^2)$. If we use the standard filtration, then grM = A/(x) and $grA/(y^2)$, so by Theorem 4.3, M · N = grM · grN = e(grM)e(grN)/e(grA) = 2. However, if we use I -adic filtration, then grM = A/(x) and grN = A/(y) and grM ·grN = e(grM)e(grN)/e(grA) = 1 ≠ M·N. This happens because F N is not discrete in I -adic filtration.

ACKNOWLEDGMENTS

Many inspirations in this research stemmed from the lecture notes of S.P. Smith [18]. I wish to thank him for his giving beautiful lectures when I was at University of Washington. I also wish to thank L.L. Avramov and C.J. Chan for many useful discussions in my preparing this paper.

REFERENCES

1. J.P. Serre, Algèbre Locale – Multiplicités, Lecture Notes in Math., Vol. 11, Springer-Verlag, Berlin, 1961.
2. P. Roberts, The vanishing of intersection multiplicities of perfect complexes, Bull. Amer. Math. Soc. 13 (1985) 127–130.
3. H. Gillet, G. Soulé, K-théorie et nullité des multiplicités d'intersection, C. R. Acad. Sci. Paris Série 1 300 (1985) 71–74.
4. C. Peskine, L. Szpiro, Syzygies et multiplicities, C. R. Acad. Sci. Paris Série A 278 (1974) 1421– 1424.
5. I. Mori, S.P. Smith, Bézout's theorem for noncommutative projective spaces, J. Pure Appl. Alg. 157 (2001) 279–299.
6. C.J. Chan, An intersection multiplicity in terms of Ext-modules, Proc. Amer. Math. Soc. 130 (2001) 327–336.
7. M. Artin, J.J. Zhang, Noncommutative projective schemes, Adv. Math. 109 (1994) 228–287.
8. P. Jorgensen, Local cohomology of non-commutative graded algebras, Comm. Algebra 25 (1997) 575–591.
9. J.J. Zhang, Connected graded Gorenstein algebras with enough normal elements, J. Algebra 189 (1997) 390–405.
10. P. Jorgensen, Noncommutative graded homological identities, J. London Math. Soc. 57 (1998) 336–350.
11. P. Jorgensen, J.J. Zhang, Gourmet's guide to Gorensteinness, Adv. Math. 151 (2000) 313–345.
12. L.L. Avramov, R. Buchweitz, Lower bounds for Betti numbers, Compositio Math. 86 (1993) 147–158. I. Mori / Journal of Algebra 252 (2002) 241–257 257
13. L.L. Avramov, R. Buchweitz, J.D. Sally, Laurent coefficients and Ext of finite graded modules, Math Ann. 307 (1997) 401–415.
14. L.L. Avramov, H.-B. Foxby, Homological dimensions of unbounded complexes, J. Pure Appl. Alg. 71 (1991) 129–155.
15. F. Ischebeck, Eine dualität zwischen den funktoren Ext und Tor, J. Algebra 11 (1969) 510–531.
16. J.C. McConnell, J.C. Robson, Noncommutative Noetherian Rings, Wiley-Interscience, Chichester, 1987.

17. C. Nastacescu, F. Van Oystaeyen, Graded and Filtered Rings and Modules, Lecture Notes in Math., Vol. 758, Springer-Verlag, Berlin, 1979.
18. S.P. Smith, Non-commutative Algebraic Geometry, Lecture notes at University of Washington, 1994.
19. C. Weibel, an Introduction to Homological Algebra, Cambridge Studies in Advanced Mathematics, Cambridge University Press, Cambridge, 1995.

CITATION

1. Izuru Mori, Intersection multiplicity over noncommutative algebras, Journal of Algebra, Volume 252, Issue 2, 15 June 2002, Pages 241-257, ISSN 0021-8693, http://dx.doi.org/10.1016/S0021-8693(02)00016-9.

Noncommutative Localisation In Algebraic *K*-Theory II

Amnon Neeman
Centre for Mathematics and Its Applications, Mathematical Sciences Institute, John Dedman Building, the Australian National University, Canberra, ACT 0200, Australia

6

ABSTRACT

In [Amnon Neeman, Andrew Ranicki, Noncommutative localisation in algebraic K-theory I, Geom. Topol. 8 (2004) 1385–1425] we proved a localisation theorem in the algebraic K-theory of noncommutative rings. The main purpose of the current article is to express the general theorem of the previous paper in a more user-friendly fashion, in a way more suitable for applications. In the process we compare our result to the existing theorems in the literature, showing how the previous paper improves all the existing results. It should be pointed out that there have been two very interesting recent preprints on related topics. The reader is referred to the beautiful papers of Krause [Henning Krause, Cohomological quotients and smashing localizations, http://wwwmath.upb.de/~hkrause/publications.html. [8]] and Dwyer [William G. Dwyer, Noncommutative localization in homotopy theory, preprint, http://www.nd.edu/~wgd/. [4]]. Krause studies the lifting of chain complexes and the relation with the telescope conjecture, and Dwyer generalises to the homotopy theoretic framework.

INTRODUCTION

This article is a sequel to [10]. We begin by briefly recalling the main result of [10]. Let A be an associative ring with unit. Let $\sigma = \{si :Pi \rightarrow Qi\}$ be a set of homomorphisms of finitely generated, projective (left) A-modules. Let $\sigma^{-1}A$ be the Cohn localisation. The main theorem of [10] is a localisation theorem; it identifies the homotopy fiber of the natural map $K(A) \rightarrow K(\sigma^{-1}A)$. Actually, only up to nonsense in degree −1. The homotopy fiber is a spectrum which, in general, has a nonvanishing (−1)th homotopy group. The main theorem of [10] says nothing about π_{-1} of the fiber. What it really gives is the (−1)-connected cover of the homotopy fiber. In order to state the theorem precisely, we first remind the reader of our notation. Let $A \rightarrow \sigma^{-1}A$ be as above. Let $C_{perf}(A)$ be the Waldhausen category of perfect complexes of A-modules. The objects are the perfect complexes, the morphisms are the chain maps, the co-fibrations are the chain maps which are split monomorphisms in each degree, and the weak equivalences are the homotopy equivalences. In $C_{perf}(A)$, let R be generated by σ . That is, R is the smallest Waldhausen subcategory of $C_{perf}(A)$ containing σ and all acyclic complexes, and closed under mapping cones and direct summands. The main K-theoretic result from [10] asserts:

Theorem 0.1: Suppose that the localisation $\sigma^{-1}A$ is stably flat over A. We remind the reader: The localisation $A \rightarrow \sigma^{-1}A$ is stably flat if, for all n>0,

$$\mathrm{Tor}_n^A\left(\sigma^{-1}A, \sigma^{-1}A\right) = 0.$$

Then the (−1)-connected cover of the homotopy fiber of $K(A) \rightarrow K(\sigma^{-1}A)$ is naturally identified with K(**R**).

This theorem gives, in great generality, a description of the (−1)-connected cover of the homotopy fiber of the map $K(A) \rightarrow K(\sigma^{-1}A)$. In practise one is often willing to sacrifice some generality to obtain a smaller, more manageable model for the homotopy fiber. The usual theorems have generally made simplifying assumptions about the set

σ = {si :Pi → Qi } of morphisms that become inverted. The usual assumption is that all the morphisms in σ are monomorphisms. Both to compare our results with older work, and more importantly in order to see what the theorem says in this very useful special case, we will study the situation that arises. If A is commutative and the maps in σ are all monomorphisms then Quillen showed (see Grayson [7, p. 229]) that the homotopy fiber of $K(A) \rightarrow K(\sigma^{-1}A)$ can be obtained as the Quillen K-theory of a relatively small and comprehensible exact category E. We generalise this to the noncommutative case. This is probably the most important, useful result of the current article.

But it seems sensible to be through about comparing our result to older work. The article does this. It should be noted that the article is purely K-theoretic. The applications to L-theory will come separately, in [11]. In this article we compare the K-theory results that can be deduced from [10] with what exists in the literature. We show quite explicitly how the methods can be used to sharply improve three existing theorems in the K-theory literature. The first is Weibel–Yao [14]. Weibel and Yao study the case where all the elements of σ are maps

$$s_i : A \rightarrow A.$$

In other words, in σ = {si :Pi →Qi } we have Pi = Qi = A for every i. The elements of σ are 1 × 1 matrices, that is ordinary elements of A. Weibel and Yao [14, p. 220] prove a localization theorem, under some extra hypotheses. The hypotheses of their theorem only hold if $\sigma^{-1}A$ is flat over A.

Let us note that if $\sigma^{-1}A$ is flat, either as a left or as a right A-module, then for all n > 0 the groups $\mathrm{Tor}_n^A(\sigma^{-1}A, \sigma^{-1}A)$ must vanish. In other words, flatness implies stable flatness. In Theorem 0.1 we see that stable flatness suffices for a localisation theorem. The first thing the reader might wish to see is examples of stably flat localisations which are not flat; this will establish that Theorem 0.1 is genuinely more powerful than Weibel and Yao [14].

We see such an example in Section 1. In the example, σ is even a set of 1 × 1 matrices. Explicitly, Section 1 will establish the easy

Example 0.2: Let k be a field, and let A = k(x, y) be the free associative algebra in two noncommuting variables x and y. Let σ consist of the single 1×1 matrix

$$x : A \longrightarrow A.$$

We assert that

(i) The ring $\sigma^{-1}A$ is not flat over A, either as a right or as a left module.
(ii) The ring $\sigma^{-1}A$ is stably flat over A.

In other words, even in the "classical" case where we only invert elements of the ring A (as opposed to matrices of them), Theorem 0.1 applies in situations not covered by Weibel and Yao [14]. For further examples, including localisations which are not even stably flat, see [12].

Next we come to the main result of the article, generalising the theorem of Quillen's. We begin by reminding the reader of Quillen's theorem; see Grayson [7, p. 229]. Grayson states the theorem in the language of algebraic geometry. If we assume that his scheme X is affine, with X = Spec(A), the theorem simplifies to

Theorem 0.3: Let A be a commutative ring, and let σ = {a} be a set containing one nonzerodivisor in A. Let T be the exact category of all finitely presented A-modules M such that

(i) There exists an integer n>0 with $a^n M = 0$.
(ii) The module M has projective dimension ≤ 1.

Then the (−1)-connected cover of the homotopy fiber of the map K(A) → $K(\sigma^{-1}A)$ is the

Quillen K-theory K(T).

The immediate question that springs to mind is what happens in the noncommutative situation. Suppose we are willing to assume that all the maps in σ are monomorphisms. We would like to define an exact category E of torsion modules for the pair (A, σ). And we would like the Quillen K-theory K(E) to agree with the (−1)-connected cover of the homotopy fiber of the map $K(A) \to K(\sigma^{-1}A)$.

Definition 0.4: Assume all the maps in σ are monomorphisms. We define an exact category ε. It is a full subcategory of the category of all A-modules. All the objects in ε are finitely presented A-modules, of projective dimension ≤ 1. The category ε is completely determined by

(i) For every $s_i : P_i \to Q_i$ in σ , the cokernel $M_i = Q_i/P_i$ lies in ε.

(ii) In any short exact sequence of finitely presented A-modules of projective dimension ≤ 1

$$0 \longrightarrow M' \longrightarrow M \longrightarrow M'' \longrightarrow 0,$$

if two of the objects M′, M and M″ lie in ε then so does the third.

(iii) ε contains all direct summands of its objects.

(iv) ε is minimal, subject to (i)–(iii).

If we want to remind ourselves that ε depends on A and σ, we will write it as ε(A, σ). If we wish to refer to it in words (as opposed to symbols), we will call it the category of (A, σ)-torsion modules. An object of ε will be an (A, σ)-torsion module.

This defined the exact category ε = ε(A, σ). To state the next theorem, we remind the reader of the notation of [10]. The Waldhausen category R has an associated derived category, which in [10] we denoted $D(\mathbf{R}) \cong Rc$. Our next result is

Theorem 0.5: Assume all the maps in σ are monomorphisms. With ε as in Definition 0.4 and $R^c \cong D(\mathbf{R})$ as in [10], there is a natural equivalence of categories $D^b(\varepsilon) \cong R^c$. The equivalence is compatible with choices of models, inducing a homotopy equivalence in K-theory

$$K(\mathcal{E}) \cong K(C^b(\mathcal{E})) \cong K$$

(Proof in Section 2.)

Remark 0.6: It should be noted that in Theorem 0.5 we do not assume that A→σ–1A is a stably flat localisation. The hypotheses are as stated in the theorem.

If $A \rightarrow \sigma^{-1}A$ happens to be stably flat, and all the maps in σ are monomorphisms, then Theorems 0.1 and 0.5 both apply. We combine our results: Theorem 0.1 tells us that K(R) is the (–1)-connected cover of the homotopy fiber of the map $K(A) \rightarrow K(\sigma^{-1}A)$. And Theorem 0.5 says that K(R) can be identified with K(ε). So far, we have identified the (–1)-connected cover of the homotopy fiber with the Quillen K-theory of some exact category ε (A, σ) of torsion modules. All that remains is to show that, in the special case where A is commutative and σ ={a} ⊂ A, our more general category ε specialises to Quillen's exact category T of Theorem 0.3.

To this end, we note

Proposition 0.7: An A-module M belongs to E if and only if

(i) M is finitely presented, and of projective dimension ≤ 1.

(ii) $\{\sigma^{-1}A\} \otimes_A M = 0 = \mathrm{Tor}_1^A(\sigma^{-1}A, M)$.

(Proof in Corollary 3.3.)

In the special case where A is commutative and σ = {a}, $\sigma^{-1}A$ is flat over A. Hence $\mathrm{Tor}_1^A(\sigma^{-1}A, M) = 0$. Since M is finitely generated, the module

$$\{\sigma^{-1}A\} \otimes_A M = A\left[\frac{1}{a}\right] \otimes_A M$$

vanishes if and only if there exists an integer $n > 0$ with $a^n M = 0$. Proposition 0.7 now tells us that E precisely agrees with Quillen's T . Hence Theorem 0.3 is a formal consequence of Theorems 0.5 and 0.1.

If $A \to \sigma^{-1}A$ is a stably flat localisation, then Theorem 0.1 gives a long exact sequence

$$\cdots \to K_2(\sigma^{-1}A) \to K_1(\mathbf{R}) \to K_1(A) \to K_1(\sigma^{-1}A) \to K_0(\mathbf{R}) \to K_0(A) \to K_0(\sigma^{-1}A)$$

If the localisation is not stably flat, the best result to date is Schofield's [13, Theorem 4.12, p. 60].

We remind the reader:

Theorem 0.8: (Schofield). Suppose the morphism $A \to \sigma^{-1}A$ is injective. Then, with the exact category E of Definition 0.4, there is an exact sequence

$$K_1(A) \longrightarrow K_1(\sigma^{-1}A) \longrightarrow K_0(\mathcal{E}) \longrightarrow K_0(A) \longrightarrow K_0(\sigma^{-1}A)$$

There is no hypothesis that $A \to \sigma^{-1}A$ is stably flat.

Remark 0.9: It should be noted that if $A \to \sigma^{-1}A$ is injective then all the maps in σ must be monomorphisms. This is well known, but we include a brief proof in Proposition 2.2. If $A \to \sigma^{-1}A$ is injective, then the category E of Definition 0.4 is well-defined. The assertion of

Theorem 0.8 makes sense.

It is natural to ask if we can prove Schofield's theorem by our techniques, or maybe even improve on it. It turns out that we can. We will prove that, with no injectivity hypothesis on maps in σ, there is always an exact sequence

$$K_1(\mathbf{R}) \to K_1(A) \to K_1(\sigma^{-1}A) \to K_0(\mathbf{R}) \to K_0(A) \to K_0(\sigma^{-1}A)$$

If the maps in σ all happen to be monomorphisms then Theorem 0.5 tells us K(ε) = K(R). We recover a longer version of Schofield's exact sequence; we have six terms, Schofield only had five. Even better: our theorem generalises to arbitrary σ.

The way our proof works is the following. In [10] we produced a diagram

$$\begin{array}{ccccccccccc} K_1(\mathbf{R}) & \longrightarrow & K_1(A) & \longrightarrow & K_1(\mathbf{T}) & \longrightarrow & K_0(\mathbf{R}) & \longrightarrow & K_0(A) & \longrightarrow & K_0(\mathbf{T}) \\ & & & & \big\downarrow {\scriptstyle K_1(T)} & & & & & & \big\downarrow {\scriptstyle K_0(T)} \\ & & & & K_1(\sigma^{-1}A) & & & & & & K_0(\sigma^{-1}A) \end{array}$$

By [10, Corollary 4.9] we know that the top row is exact. In fact, this exact sequence can be continued arbitrarily far to the left. What we prove here is:

Theorem 0.10. The maps

$$K_1(T): K_1(\mathbf{T}) \longrightarrow K_1(\sigma^{-1}A)$$
$$\textit{and} \quad K_0(T): K_0(\mathbf{T}) \longrightarrow K_0(\sigma^{-1}A)$$

are both isomorphisms.

(Proof for K0 in Section 5, for K1 in Section 6.)

Up until now we have discussed how the results of [10] can be used to improve and generalize known localisation theorems in algebraic K-theory. Our next theorem says that the results are best possible by the techniques.

We first remind ourselves a little of the proof of the main theorem of [10]. In glorious generality we constructed a diagram of triangulated categories

$$\begin{array}{ccccccc} \mathcal{R}^c & \longrightarrow & D^{\mathrm{perf}}(A) & \xrightarrow{\ \pi\ } & \mathcal{T}^c & \xrightarrow{\ T\ } & D^{\mathrm{perf}}(\sigma^{-1}A). \\ & & & \searrow \quad\quad \nearrow {\scriptstyle i} & & & \\ & & & \dfrac{D^{\mathrm{perf}}(A)}{\mathcal{R}^c} & & & \end{array}$$

The category T^c in this diagram was defined to be the idempotent completion of $\frac{D^{perf}(A)}{R^c}$. The functor i is by definition fully faithful, and every object of T^c is a direct summand of an object in the image of i. The key result [10, Theorem 0.7], was that T is an equivalence of categories if and only if the localisation is stably flat. Compacting the diagram to

$$\begin{array}{ccccc} \mathcal{R}^c & \longrightarrow & D^{\mathrm{perf}}(A) & \xrightarrow{T\pi} & D^{\mathrm{perf}}(\sigma^{-1}A) \\ & & \searrow & & \nearrow_{Ti} \\ & & & \frac{D^{\mathrm{perf}}(A)}{\mathcal{R}^c} & \end{array}$$

the key result can be restated to say that the localisation is stably flat if and only if the functor T I is fully faithful, and every object of Dperf(σ^{-1}A) is a direct summand of an object in the image of T i. The K-theoretic consequences follow formally from results of Waldhausen, Grayson and Gillet, after we make a reasonable choice of Waldhausen models. Part of the work in [10] was to establish that a suitable choice of models can be made.

Still at the level of triangulated categories, it is not a priori clear that R^c is the best possible choice. We could consider other subcategories K $\subset D^{perf}(A)$. It is conceivable that, for some other choice of K $\subset D^{perf}(A)$, the natural map

$$\frac{D^{\mathrm{perf}}(A)}{\mathcal{K}} \xrightarrow{\nu} D^{\mathrm{perf}}(\sigma^{-1}A)$$

will be fully faithful, and that every object in $D^{perf}(\sigma^{-1}A)$ will be isomorphic to a direct summand of something in the image of v. Our next theorem asserts that the only choice for K which has any chance of working is K = R^c. As we already know, K = R^c only works when the localization is stably flat.

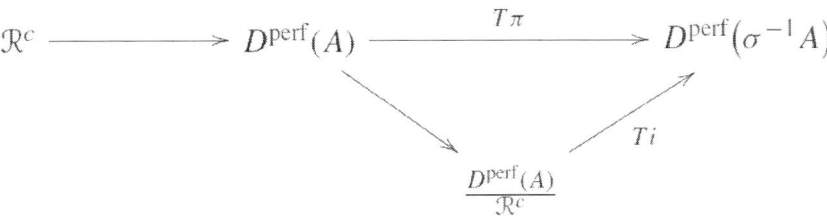

If the functor

$$\frac{D^{\mathrm{perf}}(A)}{\mathcal{K}} \xrightarrow{\nu} D^{\mathrm{perf}}\left(\sigma^{-1}A\right)$$

is to be fully faithful, then the objects that map to zero by the composite

$$D^{\mathrm{perf}}(A) \xrightarrow{\mu} \frac{D^{\mathrm{perf}}(A)}{\mathcal{K}} \xrightarrow{\nu} D^{\mathrm{perf}}\left(\sigma^{-1}A\right)$$

are precisely those in $K \subset$ Dperf(A). This is only of interest if the composite $\nu\mu$ induces the natural map $K(A) \rightarrow K(\sigma^{-1}A)$. The natural functor that does this is $T\pi : D^{\mathrm{perf}}(A) \rightarrow D^{\mathrm{perf}}(\sigma^{-1}A)$. We remind the reader that $T\pi$ is the functor taking X to $\{\sigma^{-1}A\}^{L} \otimes_{A} X$. To show that $K = R^{c}$ is the only possibility, we prove

Theorem 0.11: Let X be an object of Dperf(A). Then $T\pi(X) \cong 0$ if and only if $X \in R^{c}$. In the notation above $K = R^{c}$ is the only possible choice. Our K-theoretic results are optimal by the methods.

(Proof in Proposition 3.2.)

This ends the summary of the theorems in the article. The results are mostly independent of each other, and the order in which the proofs are presented is a little arbitrary. We did try to put near the beginning the results which are relevant to the next article in this series, about L-theory. The results of Sections 5 and 6 play no role in the L-theory work. For this reason they are at the end. The technical lemmas of Section 4 are irrelevant to Section 5 but are needed in Section 6. From a logical point of view it might have been better to put them between the two sections. But, since these technical lemmas also play a role in L-theory, we placed them before the last two sections. This way an L-theorist can safely skip the last two sections of the article.

The comparison with Weibel–Yao

In this section we will establish the very easy Example 0.2. We remind the reader. Let k be any field. The ring A is the free associative algebra k_x,y_ in two variables x and y. The set σ contains the singleton {x :A→A}. We wish to prove that

(i) $\sigma^{-1}A$ is not flat over A, either as a right or as a left A-module.
(ii) $\sigma^{-1}A$ is stably flat over A.

Let us prove (ii) first.We know from [3, Corollary 2, p. 68] that the ring A = k(x, y) is hereditary; that is, every module has projective dimension ≤ 1. It follows that $\mathrm{Tor}_n^A(\sigma^{-1}A, \sigma^{-1}A)$ must vanish whenever n > 1. The fact that $\mathrm{Tor}_n^A(\sigma^{-1}A, \sigma^{-1}A)$ vanishes may be found in Schofield [13, p. 58], or in [10, Lemma 8.6(ii)]. Putting this together we have that $\mathrm{Tor}_n^A(\sigma^{-1}A, \sigma^{-1}A)$ vanishes for all n ≤ 1, that is the localisation is stably flat.

Next we prove (i). By symmetry it suffices to show that $\sigma^{-1}A$ is not flat as a right A-module. Put k = A/(Ax + Ay). Since the intersection of the left ideals Ax and Ay is trivial, we have a free resolution for k

$$0 \longrightarrow A \oplus A \xrightarrow{(x,y)} A \longrightarrow k \longrightarrow 0.$$

Tensoring this with $\sigma^{-1}A$ we deduce an exact sequence

$$0 \longrightarrow \mathrm{Tor}_1^A(\sigma^{-1}A, k) \longrightarrow \sigma^{-1}A \oplus \sigma^{-1}A \xrightarrow{(x,y)} \sigma^{-1}A \longrightarrow \sigma^{-1}A \otimes_A k \longrightarrow 0.$$

By the choice of σ, the map x : $\sigma^{-1}A \to \sigma^{-1}A$ is an isomorphism. The kernel of the map (x, y) : $\sigma^{-1}A \oplus \sigma^{-1}A \to \sigma^{-1}A$ therefore identifies as $\sigma^{-1}A$. Thus

$$\mathrm{Tor}_1^A(\sigma^{-1}A, k) = \sigma^{-1}A,$$

and $\sigma^{-1}A$ is not flat as a right A-module.

TORSION MODULES

Hypothesis 2.1: In this section, we assume that all the morphisms in σ are injections.

The following result is well known; since the proof is so easy we include it.

Proposition 2.2: If $A \to \sigma^{-1}A$ is an injection then every si $:P_i \to Q_i$ in σ is an injection, i.e.

Hypothesis 2.1 is satisfied.

Proof: Since $A \to \sigma^{-1}A$ is a monomorphism and P_i is projective and therefore flat, we deduce that

$$A \otimes_A P_i \longrightarrow \{\sigma^{-1}A\} \otimes_A P_i$$

is a monomorphism. Abbreviate $\{\sigma^{-1}A\} \otimes_A P_i$ as $\sigma^{-1}P_i$. Then the above says that the map $P_i \to \sigma^{-1}$Pi is mono. Consider the commutative diagram

$$\begin{array}{ccc} P_i & \xrightarrow{\ s_i\ } & Q_i \\ \downarrow & & \downarrow \\ \sigma^{-1}P_i & \xrightarrow{\ \sigma^{-1}s_i\ } & \sigma^{-1}Q_i \end{array}$$

By the above, Pi $\to\sigma-1$Pi is an injection. Now $\sigma^{-1}s_i : \sigma^{-1}P_i \to \sigma^{-1}Q_i$ is an isomorphism.

From the commutativity of the square we deduce that si $:P^i \to Q_i$ is an injection.

Example 2.3: The converse of Proposition 2.2 does not hold in general. The set $\sigma = \{0 \to A\}$ satisfies Hypothesis 2.1, but $A \to \sigma^{-1}A = 0$ is not injective.

In the rest of this section we will always assume that Hypothesis 2.1 holds. That is, all the maps $s_i : P_i \to Q_i$ are injective. Then all the cokernels $M_i = Q_i/P_i$ fit in short exact sequences

$$0 \longrightarrow P_i \xrightarrow{s_i} Q_i \longrightarrow M_i \longrightarrow 0.$$

The modules M_i are all finitely presented, and all have projective dimension ≤ 1. In Definition 0.4 we let ε = ε(A, σ) be the smallest exact subcategory of the category of all finitely presented A-modules of projective dimension ≤ 1, which contains the M_i above and is closed under short exact sequences and direct summands. We remind the reader

Definition 2.4: The bounded derived category of the exact category ε, denoted $D^b(\varepsilon)$, is defined as follows. The objects are bounded chain complexes of objects of E. The morphisms are obtained from the chain maps by formally inverting the maps whose mapping cones are acyclic (as complexes of A-modules). There is an obvious functor i :$D^b(\varepsilon) \to D(A)$.

Lemma 2.5: The functor i :$D^b(\varepsilon) \to D(A)$ is fully faithful.

Proof: Let us begin by showing that, for any objects M,N ∈ ε and any $n \in \mathbb{Z}$,

$$\{D^b(\mathcal{E})\}(M, \Sigma^n N) = \{D(A)\}(M, \Sigma^n N)$$

Take a map in $D^b(\varepsilon)$ of the form $M \to \Sigma^n N$. There exists a bounded complex X of objects in ε, a quasi-isomorphism g :X→M, and a map of complexes f :$X \to \Sigma^n N$ so that our map is fg^{-1}. That is, we have a complex

$$\cdots \longrightarrow X^{-2} \longrightarrow X^{-1} \longrightarrow X^0 \xrightarrow{\partial_0} X^1 \longrightarrow \cdots.$$

There is a quasi-isomorphism X→M; in particular $H^0(X) = M$. We have an exact sequence

$$\longrightarrow X^{-1} \longrightarrow \ker(\partial_0) \longrightarrow M \longrightarrow 0$$

But M ∈ ε means that M is of projective dimension ≤ 1. There is an exact sequence

$$0 \longrightarrow P \xrightarrow{s} Q \longrightarrow M \longrightarrow 0$$

with P and Q (finitely generated) and projective A-modules. Since P and Q are projective Amodules, there exists a map

$$\begin{array}{ccccccc} P & \longrightarrow & Q & \longrightarrow & M & \longrightarrow & 0 \\ \downarrow & & \downarrow & & \downarrow{\scriptstyle 1} & & \\ X^{-1} & \longrightarrow & \ker(\partial_0) & \longrightarrow & M & \longrightarrow & 0. \end{array}$$

Let Z be given by the pushout square

$$\begin{array}{ccc} P & \longrightarrow & Q \\ \downarrow & & \downarrow \\ X^{-1} & \longrightarrow & Z. \end{array}$$

In the short exact sequence

$$0 \longrightarrow X^{-1} \longrightarrow Z \longrightarrow M \longrightarrow 0$$

we are given that X^{-1} and M lie in ε. It immediately follows that Z is finitely presented and of projective dimension ≤ 1. Definition 0.4(ii) now establishes that Z ∈ ε. The short exact sequence also tells us that the complex $0 \to X^{-1} \to Z \to 0$ is quasi-isomorphic to M. We deduce a quasi-isomorphism h : X→X of complexes, given below:

$$\begin{array}{ccccccccccc} \cdots & \longrightarrow & 0 & \longrightarrow & X^{-1} & \longrightarrow & Z & \longrightarrow & 0 & \longrightarrow & \cdots \\ & & \downarrow & & \downarrow & & \downarrow & & \downarrow & & \\ \cdots & \longrightarrow & X^{-2} & \longrightarrow & X^{-1} & \longrightarrow & X^{0} & \longrightarrow & X^{1} & \longrightarrow & \cdots. \end{array}$$

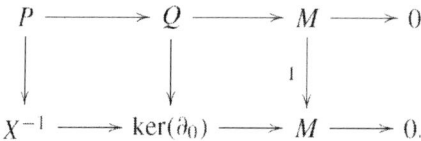

It follows that the map $fg^{-1} : M \to \Sigma^n N$ is equal to the map $\{fh\}\{gh\}^{-1}$. Since X′ is concentrated in degrees $^{-1}$ and 0, it follows that fh vanishes unless n = 0 or 1. Unless n = 0 or 1, we have proved that $\{D^b(\varepsilon)\}(M,\Sigma^n N)$ vanishes. As for

$$\{D(A)\}(M, \Sigma^n N) = \mathrm{Ext}^n_A(M, N)$$

it must vanish since the projective dimension of M is ≤ 1. In other words, for n ≠ 0,1 the equality

$$\{D^b(\mathcal{E})\}(M, \Sigma^n N) = \{D(A)\}(M, \Sigma^n N)$$

is just because both sides vanish.

We leave to the reader to check that the two sides are equal also when n = 0 or 1. For n = 0 both sides identify as $\varepsilon(M,N)$, while for n = 1 both sides identify as $\mathrm{Ext}^1_A(M,N)$.

Let M be an object of ε. Consider next the full subcategory $B \subset D^b(\varepsilon)$ defined by

$$\mathrm{Ob}(\mathcal{B}) = \{Y \in \mathrm{Ob}(D^b(\mathcal{E})) \mid \\ \forall n \in \mathbb{Z},\ \{D^b(\mathcal{E})\}(M, \Sigma^n Y) \to \{D(A)\}(M, \Sigma^n Y) \text{ is an isomorphism}\}$$

By the above B contains ε, and clearly B is triangulated. Hence B contains all of $D^b(\varepsilon)$. Next, take any Y in $D^b(\varepsilon)$, and consider the full subcategory $C \subset D^b(\varepsilon)$ given by

$$\mathrm{Ob}(\mathcal{C}) = \{X \in \mathrm{Ob}(D^b(\mathcal{E})) \mid \\ \forall n \in \mathbb{Z},\ \{D^b(\mathcal{E})\}(X, \Sigma^n Y) \to \{D(A)\}(X, \Sigma^n Y) \text{ is an isomorphism}\}$$

By the above $\varepsilon \subset C$, and C is clearly triangulated. Hence C contains Db(ε).

Lemma 2.6: Assume that maps in σ are all injections. The natural map $D^b(\varepsilon) \to D(A)$ factors through $R^c \subset D(A)$, and the induced map $D^b(\varepsilon) \to R^c$ is an equivalence of categories.

Proof: Let ε be the full subcategory of S = D(A) containing all objects isomorphic to objects in the image of i :$D^b(\varepsilon) \to D(A)$. By Lemma 2.5, ε is a full, triangulated subcategory of S = D(A). If M_i fits in an exact sequence

$$0 \longrightarrow P_i \xrightarrow{s_i} Q_i \longrightarrow M_i \longrightarrow 0$$

with $s_i \in \sigma$, then $M_i \in \varepsilon \subset D^b(\varepsilon)$. Thus M_i, which is quasi-isomorphic to the complex

$$\cdots \longrightarrow 0 \longrightarrow P_i \xrightarrow{s_i} Q_i \longrightarrow 0 \longrightarrow \cdots$$

lies in $D^b(\varepsilon) \subset \varepsilon \subset D(A)$. Since E contains any isomorph of any of its objects, σ is contained in $\varepsilon \subset D(A)$.

From [9, Lemma 2.2] we know that $R^c \subset D(A)$ is the smallest thick subcategory containing σ . That is, R^c is the smallest triangulated subcategory of D(A), containing σ and closed under direct summands. We know that $D^b(\varepsilon)$ is equivalent to ε, and ε contains σ . We want to show that $\varepsilon = R^c$. All we need for this is to prove that ε is closed under direct summands and minimal. The minimality is clear, from the minimality of ε. We need to show that ε is closed under direct summands.

But Definition 0.4(iii) tells us that every idempotent in ε splits. Theorem 2.8 of Balmer and Schlichting's [1] allows us to deduce that $D^b(\varepsilon)$ is idempotent complete; all direct summands of objects in $D^b(\varepsilon)$ lie in $D^b(\varepsilon)$ (up to isomorphism). Hence the equivalent category E is also closed under direct summands.

Theorem 2.7: Suppose every morphism in σ is injective. Then the Waldhausen K-theory of the Waldhausen category R is isomorphic to the Quillen K-theory of the exact category ε.

Proof: By Lemma 2.6, the natural map $D^b(\varepsilon) \to D(A)$ induces a triangulated equivalence of $D^b(\varepsilon)$ with R^c. In order to turn this into a statement in K-theory we need to choose models wisely.

Let C′(A) be the followingWaldhausen model category. The objects are bounded chain complexes of A-modules, where every module in the chain complex is finitely presented and of projective dimension ≥ 1. The morphisms are the chain maps. The cofibrations are the degreewise split monomorphisms, and the weak equivalences are the homology isomorphisms. There is an inclusion functor of Waldhausen categories

$$C^{\text{perf}}(A) \longrightarrow C'(A)$$

On the level of derived categories it induces an equivalence $D^{\text{perf}}(A) \to D'(A)$. The inclusion $R \to C^{\text{perf}}(A)$ induces a fully faithful triangulated functor $D(R) \to D^{\text{perf}}(A) \simeq D'(A)$.

We define $R' \subset C'(A)$ to be the full Waldhausen subcategory of all objects which become isomorphic in D′(A) to objects in the image of $D(\mathbf{R}) \simeq \mathfrak{R}^c$. By [10, Lemma 2.5], the natural map D(R)→D(R′) is an equivalence.

Let $C^b(\varepsilon)$ be the following Waldhausen model category. The objects are bounded chain complexes of objects of ε. The morphisms are the chain maps. The cofibrations are the degreewise split monomorphisms, and the weak equivalences are the homology isomorphisms. The natural inclusions give maps of Waldhausen model categories

$$\begin{array}{ccc} & & C^b(\mathcal{E}) \\ & & \downarrow \\ \mathbf{R} & \longrightarrow & \mathbf{R}'. \end{array}$$

At the level of derived categories we get equivalences

$$\begin{array}{ccc} & & D^b(\mathcal{E}) \\ & & \downarrow \wr \\ D(\mathbf{R}) & \xrightarrow[\ -\ ]{\sim} & D(\mathbf{R}'). \end{array}$$

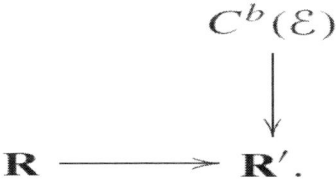

By [10, Theorem 2.2] we deduce homotopy equivalences of K-theory spectra

$$\begin{array}{ccc} & & K(C^b(\mathcal{E})) \\ & & \downarrow \wr \\ K(\mathbf{R}) & \xrightarrow{\sim} & K(\mathbf{R}'). \end{array}$$

Now Gillet's [5, 6.2] tells us that the natural map $K(\varepsilon) \to K(C^b(\varepsilon))$ is a homotopy equivalence, completing the construction of a homotopy equivalence $K(\varepsilon) \simeq K(R)$.

THE KERNEL OF THE TENSOR PRODUCT

The main aim of this section is to prove Theorem 0.11, which we will do in Proposition 3.2. As a corollary we will also deduce Proposition 0.7. But first it might help to remind the reader of our notation.

Recall [10, Lemma 5.3]. The tensor product defines a functor $D(A) \to D(\sigma^{-1}A)$. It is the functor

$$X \mapsto \{\sigma^{-1}A\}^L \otimes_A X$$

In [10, Lemma 5.3] we proved that this functor has a natural factorisation

$$D(A) \xrightarrow{\pi} \mathcal{T} \xrightarrow{T} D(\sigma^{-1}A)$$

Furthermore, we proved that the functor T takes compacts to compacts. We have induced functors

$$D^{\text{perf}}(A) \xrightarrow{\pi} \mathcal{T}^c \xrightarrow{T} D^{\text{perf}}(\sigma^{-1}A)$$

The main results of this section are the following statements, which are slight variants of each other.

(i) Let X be an object of $D^{perf}(A)$. We have $T\pi(X) \cong 0$ if and only if $X \in R^c$. (See Proposition 3.2.)

(ii) Let t be an object of T^c. If $T(t) \cong 0$ then $t \cong 0$. (See Proposition 3.4.)

For the sake of compactness of notation, we will adopt the abbreviation

$$\sigma^{-1}P = \{\sigma^{-1}A\} \otimes_A P$$

whenever P is a projective A-module. As in [10], we abbreviate $D^{perf}(A) = S^c$. And we introduce the new shorthand, for this section,

$$\mathcal{D} = D(\sigma^{-1}A)$$

Proposition 3.1: For any two finitely generated projective A-modules P and Q one has

$$\mathcal{T}^c(\pi P, \pi Q) = \mathcal{D}(T\pi P, T\pi Q) = \mathrm{Hom}_{\sigma^{-1}A}(\sigma^{-1}P, \sigma^{-1}Q)$$

Proof: The identity $D(T\pi P, T\pi Q) = \mathrm{Hom}_{\sigma^{-1}A}(\sigma^{-1}P, \sigma^{-1}Q)$ is almost by definition. After all, $T\pi$ is the functor taking P to $\sigma^{-1}P = \{\sigma^{-1}A\} \otimes_A P$. And the inclusion of the category of $\sigma^{-1}A$-modules into its derived category $D = D(\sigma^{-1}A)$ is fully faithful;

$$\mathcal{D}(T\pi P, T\pi Q) = \mathcal{D}(\sigma^{-1}P, \sigma^{-1}Q) = \mathrm{Hom}_{\sigma^{-1}A}(\sigma^{-1}P, \sigma^{-1}Q)$$

There is a natural map, induced by the functor T ,

$$\mathcal{T}^c(\pi P, \pi Q) \longrightarrow \mathcal{D}(T\pi P, T\pi Q)$$

We need to prove it an isomorphism. The case where P =Q= A is easy; we have

$$\begin{aligned}\mathcal{T}^c(\pi A,\pi A) &= \{\sigma^{-1}A\}^{op} \qquad \text{by [10, Theorem 7.4]}\\ &= \mathcal{D}(\sigma^{-1}A,\sigma^{-1}A)\\ &= \mathcal{D}(T\pi A, T\pi A).\end{aligned}$$

But the collection of all P and Q for which the map $T : T^c(\pi P, \pi Q) \to D(T\pi P, T\pi Q)$ is an isomorphism is clearly closed under finite direct sums and direct summands, and hence contains all the finitely generated projective modules.

Proposition 3.2: For any object X in $D^{perf}(A)$, we have the implication

$$\{\{\sigma^{-1}A\}^L \otimes_R X = 0\} \quad \Leftrightarrow \quad \{X \in \mathcal{R}^c\}$$

Proof: If $X \in Rc$ then $\pi X = 0$, hence $\{\sigma^{-1}A\}^L \otimes_A X = T\pi X = 0$. The non-trivial statement is the converse. We need to prove that if $T\pi X = 0$ then $X \in R^c$.

Take any $X \in D^{perf}(A)$. That is, X is a bounded complex of finitely generated projective A-modules. Up to suspension, X may be written as a complex

$$\to 0 \to X^0 \to X^1 \to \cdots \to X^{n-1} \to X^n \to 0 \to$$

and all the Xi are finitely generated and projective. Assume $\{\sigma^{-1}A\}^L \otimes_R X = 0$. We need to prove that $X \in R^c$. But $\{\sigma^{-1}A\}^L \otimes_R X = 0$ means that the complex

$$\to 0 \to \sigma^{-1}X^0 \to \sigma^{-1}X^1 \to \cdots \to \sigma^{-1}X^{n-1} \to \sigma^{-1}X^n \to 0 \to$$

must be contractible. There are maps $\sigma^{-1}X^i \to \sigma^{-1}X^{i-1}$ so that, for each i, the sum of the two composites

$$\begin{array}{ccc}
\sigma^{-1}X^{i} & \longrightarrow & \sigma^{-1}X^{i+1} \\
\downarrow & & \downarrow \\
\sigma^{-1}X^{i-1} & \longrightarrow & \sigma^{-1}X^{i}
\end{array}$$

is the identity on $\sigma^{-1}X^{i}$. By Proposition 3.1, the contracting homotopy may be lifted to the complex

$$\to 0 \to \pi X^{0} \xrightarrow{\partial} \pi X^{1} \xrightarrow{\partial} \cdots \xrightarrow{\partial} \pi X^{n-1} \xrightarrow{\partial} \pi X^{n} \to 0 \to$$

For each i there are maps $D: \pi X^{i} \to \pi X^{i-1}$, so that the two composites

$$\begin{array}{ccc}
\pi X^{i} & \xrightarrow{\partial} & \pi X^{i+1} \\
{\scriptstyle D}\downarrow & & \downarrow{\scriptstyle D} \\
\pi X^{i-1} & \xrightarrow{\partial} & \pi X^{i}
\end{array}$$

add to the identity on πX^{i}.

Now let $Y^{i} \in D^{perf}(A)$ be the complex

$$\to 0 \to X^{0} \to X^{1} \to \cdots \to X^{i-1} \to X^{i} \to 0 \to$$

For each i, there is a triangle

$$\Sigma^{-i-1}X^{i+1} \longrightarrow Y^{i+1} \longrightarrow Y^{i} \longrightarrow \Sigma^{-i}X^{i+1}$$

The functor π is triangulated, and hence for each i we deduce a triangle

$$\Sigma^{-i-1}\pi X^{i+1} \longrightarrow \pi Y^{i+1} \longrightarrow \pi Y^{i} \xrightarrow{\rho_i} \Sigma^{-i}\pi X^{i+1}$$

We shall prove, by induction on i, that

(i) The map

$$\pi Y^i \xrightarrow{\rho_i} \Sigma^{-i}\pi X^{i+1}$$

is a split monomorphism in T^c.

(ii) For each i we shall produce an explicit splitting; that is, we shall produce a map

$$\Sigma^{-i}\pi X^{i+1} \xrightarrow{\theta_i} \pi Y^i$$

so that $\vartheta^i\rho^i$ is the identity on πY^i .

(iii) $1-\rho_i\vartheta_i$ is an endomorphism of $\Sigma^{-i}\pi X^{i+1}$. We shall show it to be the composite

$$\Sigma^{-i}\pi X^{i+1} \xrightarrow{\Sigma^{-i}\partial} \Sigma^{-i}\pi X^{i+2} \xrightarrow{\Sigma^{-i}D} \Sigma^{-i}\pi X^{i+1}$$

with ∂ and D as above, satisfying $1 = D\partial + \partial D$.

Note that for $i < -1$, $X^{i+1} = Y^i = 0$, and there is nothing to do. We may assume that (i)–(iii) hold for some i. We only need to show the induction step; that is, if it holds for i then it holds also for i + 1.

It is easy to compute, in the derived category D(A), the composite $\alpha\beta$, with α and β the morphisms in the triangles below

$$\Sigma^{-i-1}X^{i+1} \xrightarrow{\beta} Y^{i+1} \longrightarrow Y^i \longrightarrow \Sigma^{-i}X^{i+1},$$

$$\Sigma^{-i-2}X^{i+2} \longrightarrow Y^{i+2} \longrightarrow Y^{i+1} \xrightarrow{\alpha} \Sigma^{-i-1}X^{i+2}$$

The morphism $\alpha\beta$ is just Σ^{-i-1} applied to the differential $\partial : X^{i+1} \to X^{i+2}$. Applying the functor π we conclude the following. By the part (iii) of the induction hypothesis, the composite

$$\Sigma^{-i-1}\pi X^{i+1} \xrightarrow{\Sigma^{-i-1}\partial} \Sigma^{-i-1}\pi X^{i+2} \xrightarrow{\Sigma^{-i-1}D} \Sigma^{-i-1}\pi X^{i+1}$$

is equal to $1-\Sigma^{-1}(\rho^i\vartheta^i)$. By the above, it factors further as

$$\Sigma^{-i-1}\pi X^{i+1} \xrightarrow{\pi\beta} \pi Y^{i+1} \xrightarrow{\pi\alpha} \Sigma^{-i-1}\pi X^{i+2} \xrightarrow{\Sigma^{-i-1}D} \Sigma^{-i-1}\pi X^{i+1}$$

Now look at the longer composite

$$\Sigma^{-i-1}\pi X^{i+1} \xrightarrow{\pi\beta} \pi Y^{i+1} \xrightarrow{\{\Sigma^{-i-1}D\}\circ(\pi\alpha)} \Sigma^{-i-1}\pi X^{i+1} \xrightarrow{\pi\beta} \pi Y^{i+1}$$

It is equal to $(\pi\beta)[1 - \Sigma^{-1}(\rho^i\vartheta^i)]$. The distinguished triangle

$$\Sigma^{-i-1}\pi X^{i+1} \xrightarrow{\pi\beta} \pi Y^{i+1} \longrightarrow \pi Y^i \xrightarrow{\rho_i} \Sigma^{-i}\pi X^{i+1},$$

coupled with the fact that ρ_i is a split monomorphism, guarantees that the triangle is really a split exact sequence in $\mathcal{T}$

$$0 \longrightarrow \Sigma^{-1}\pi Y^i \xrightarrow{\Sigma^{-1}\rho_i} \Sigma^{-i-1}\pi X^{i+1} \xrightarrow{\pi\beta} \pi\Sigma Y^{i+1} \longrightarrow 0$$

But then $\pi\beta$ is a split epimorphism, and its composite with $\Sigma^{-1}\rho_i$ vanishes. From the vanishing of $\{\pi\beta\}\{\Sigma^{-1}\rho_i\}$ it follows that

$$(\pi\beta)\left\{\Sigma^{-i-1}D\right\}(\pi\alpha)(\pi\beta) = (\pi\beta)\left[1 - \Sigma^{-1}(\rho_i\theta_i)\right] = \pi\beta.$$

Hence

$$\left[1 - (\pi\beta)\circ\left\{\Sigma^{-i-1}D\right\}\circ(\pi\alpha)\right](\pi\beta) = 0.$$

Since $\pi\beta$ is a split epimorphism, we conclude that

$$(\pi\beta)\circ\left\{\Sigma^{-i-1}D\right\}\circ(\pi\alpha) = 1.$$

But $\pi\alpha : \pi Y^{i+1} \to \Sigma^{-i-1}\pi X^{i+2}$ is nothing other than the map ρ_{i+1}, and if we put $\vartheta_{i+1} = (\pi\beta)\circ\{\Sigma^{-i-1}D\}$, then we have proved parts (i) and (ii) for i +1.

It only remains to establish (iii). By construction, $\rho_{i+1}\vartheta_{i+1}$ is given by the composite

$$\Sigma^{-i-1}\pi X^{i+2} \xrightarrow{\Sigma^{-i-1}D} \Sigma^{-i-1}\pi X^{i+1} \xrightarrow{\pi\beta} \pi Y^{i+1} \xrightarrow{\pi\alpha} \Sigma^{-i-1}\pi X^{i+2},$$

which is nothing other than Σ–i–1(∂D). Hence this equals 1 –Σ–i–1(D∂).

This completes the induction. Now choose i > n. The complex Y^i is nothing other than X $\in D^{perf}(A)$, and by (i) we conclude that πX is a direct summand of $\pi X^{i+1} = 0$. It follows that $\pi X = 0$, as an object of $\mathcal{S}^c/\mathcal{R}^c \subset \mathcal{T}^c$. This forces $X \in \mathcal{R}^c$.

Corollary 3.3: Suppose all the morphisms in σ are monomorphisms, so that the exact category E of Definition 0.4 makes sense. If M is an A-module, then M lies in E if and only if

1. M is finitely presented, and of projective dimension ≤ 1.

2. $\{\sigma^{-1}A\} \otimes_A M = 0 = \mathrm{Tor}_1^A(\sigma^{-1}A, M)$

Proof: By Proposition 3.2, $\{\sigma^{-1}A\}|^L \otimes_A X = 0$ if and only if $X \in \mathcal{R}^c$. Since all the morphisms in σ are monomorphisms, Lemma 2.6 asserts that Rc Db(ε). Combining the results, $\{\sigma^{-1}A\}^L \otimes_A X = 0$ if and only if X is isomorphic to an object in $D^b(\varepsilon)$.

If M is an object of $\varepsilon \subset D^b(\varepsilon)$, then by Definition 0.4 M is finitely presented and of projective dimension ≤ 1, and by the above $\{\sigma^{-1}A\}^L \otimes_A M = 0$. That is, $\{\sigma^{-1}A\} \otimes_A M = 0 = \mathrm{Tor}_1^A(\sigma^{-1}A, M)$. We need to prove the converse.

Suppose therefore that M is a finitely presented A-module of projective dimension ≤ 1, and that $\{\sigma^{-1}A\} \otimes_A M = 0 = \mathrm{Tor}_1^A(\sigma^{-1}A, M)$. We need to show that $M \in \varepsilon$. Because M is of

projective dimension _ 1, we have $\mathrm{Tor}_n^A(\sigma^{-1}A, M) = 0$ for all n>1. By hypothesis $\{\sigma^{-1}A\} \otimes_A M = 0 = \mathrm{Tor}_1^A(\sigma^{-1}A, M)$. It follows that $\{\sigma^{-1}A\} \otimes_A M = 0$. By the first paragraph of the proof we deduce that M is quasi-isomorphic to an object in $D^b(\varepsilon)$. There is a chain complex in ε

$$0 \to X^{-m} \to \cdots \to X^{-1} \to X^0 \to X^1 \to \cdots \to X^n \to 0$$

whose only cohomology is M in degree 0. For i ≥ 0 define K^i to be the kernel of $X^i \to X^{i+1}$. We know that $K^n = X^n \in \varepsilon$. For each i ≥ 0 we have short exact sequences

$$0 \longrightarrow K^i \longrightarrow X^i \longrightarrow K^{i+1} \longrightarrow 0.$$

If $K^{i+1} \in \varepsilon$, then we reason that both K^{i+1} and X^i lie in ε, hence are finitely presented and of projective dimension ≤ 1. The exact sequence says first that K^i is finitely presented and of projective dimension ≤ 1. But then Definition 0.4(ii) tells us that $K^i \in \varepsilon$. By descending induction we conclude $K^i \in \varepsilon$ for all i ≥s 0. In particular $K^0 \in \varepsilon$. For all i ≥ 0 define I^{-i} to be the image of $X^{-i-1} \to X^{-i}$. We have a short exact sequence

$$0 \longrightarrow I^0 \longrightarrow K^0 \longrightarrow M \longrightarrow 0.$$

We know that M is finitely presented and of projective dimension ≤ 1, while $K^0 \in \varepsilon$ (hence also finitely presented and of projective dimension ≤ 1). It follows that I 0 is finitely presented and of projective dimension ≤ 1. The short exact sequences

$$0 \longrightarrow I^{-i} \longrightarrow X^{-i} \longrightarrow I^{-i+1} \longrightarrow 0$$

tell us that if I^{-i+1} is finitely presented and of projective dimension ≤ 1, then so is I^{-i} (because $X^{-i} \in \varepsilon$ must have the property). By descending induction all the modules I^{-i} must be finitely presented and of projective dimension ≤ 1. But $I^{-m} = 0 \in \varepsilon$. The exact sequences

$$0 \longrightarrow I^{-i} \longrightarrow X^{-i} \longrightarrow I^{-i+1} \longrightarrow 0,$$

coupled with Definition 0.4(ii), tell us that if $I^{-i} \in \varepsilon$ then $I^{-i+1} \in \varepsilon$. Ascending induction allows us to conclude that all the $I^{-i} \in E$. In particular $I^0 \in \varepsilon$. But now in the exact sequence

$$0 \longrightarrow I^0 \longrightarrow K^0 \longrightarrow M \longrightarrow 0$$

we have I^0 and K^0 both lying in ε. Hence $M \in \varepsilon$.

The following is a technical improvement of Proposition 3.2 which we will need in Section 5.

Proposition 3.4: Suppose t is an object in $\mathcal{T}^c$. If Tt = 0 then t = 0.

Proof: By Proposition 3.2 we know that if $x \in D^{perf}(A)$, and if

$$T\pi x = \{\sigma^{-1}A\}^L \otimes_R x = 0,$$

then $x \in \mathcal{R}^c$, in other words $\pi x = 0$. The proposition is therefore true for all objects $\pi x \in \mathcal{T}^c$, with $x \in D^{perf}(A)$.

Now we turn to the general case. Suppose t is an object in $\mathcal{T}^c$ with T t = 0. We wish to show t = 0. By the last sentences of [9, Theorem 2.1], t is a direct summand of an object πx, with $x \in D^{perf}(A)$. That is, there exists an object $t' \in \mathcal{T}^c$ with

$$\pi x \cong t \oplus t'$$

Consider the distinguished triangle

$$t \oplus t' \xrightarrow{\begin{pmatrix} 0 & 0 \\ 0 & 1 \end{pmatrix}} t \oplus t' \longrightarrow t \oplus \Sigma t \longrightarrow \Sigma(t \oplus t').$$

It is a triangle of the form

$$\pi x \longrightarrow \pi x \longrightarrow t \oplus \Sigma t \longrightarrow \Sigma \pi x.$$

Since the image of the functor π, that is the subcategory $\mathcal{S}^c/\mathcal{R}^c \subset \mathcal{T}^c$, is full and triangulated (see [9, Theorem 2.1]), the entire distinguished triangle above must lie in $\mathcal{S}^c/\mathcal{R}^c$. That means $t \oplus \Sigma t$ is isomorphic to an object in $\mathcal{S}^c/\mathcal{R}^c$. It follows there exists an object $y \in D^{perf}(A) \cong \mathcal{S}^c$ with

$$\pi y \cong t \oplus \Sigma t$$

This makes

$$T\pi y \cong Tt \oplus \Sigma Tt \cong 0.$$

By the above $\pi y = 0$. But t is a direct summand of πy; hence t = 0.

A BOUND ON THE LENGTH OF COMPLEXES IN $\mathcal{R}^c$

The category $\mathcal{R}^c$ is the smallest thick subcategory of $D^{perf}(A)$ containing σ. Every object in $\mathcal{R}^c$ is a direct summand of an object made up of iterated mapping cones on objects in σ. For technical reasons we want bounds on how long the iterated mapping cones need to be. This section is devoted to proving such bounds.

Definition 4.1: The full subcategory of all objects in $\mathcal{S}$ = D(A) which vanish outside the range [m,n] will be denoted S[m,n]. We allow m or n to be infinite; the categories $\mathcal{S}[m,\infty)$ and $\mathcal{S}(-\infty,n]$ have the obvious definitions.

Remark 4.2: The reader should note that the categories $\mathcal{S}[n,\infty)$ and $\mathcal{S}(-\infty,n]$ should not be confused with $\mathcal{S}^{\geq n}$ and $\mathcal{S}^{\leq n}$. It is true that every object in $\mathcal{S}^{\leq n}$ is isomorphic in S to a chain complex

$$\cdots \to X^m \to X^{m+1} \to \cdots \to X^{n-1} \to X^n \to 0 \to 0 \to \cdots.$$

An isomorphism in $\mathcal{S}$ = D(A) is after all just a homology isomorphism. For any object in $\mathcal{S}^{\leq n}$, there is an object in $\mathcal{S}(-\infty,n]$ homology isomorphic to it. But for once we want to have a name for the complexes which are actually supported on the interval [m,n], not just isomorphic in $\mathcal{S}$ to such objects.

Definition 4.3: The category $\mathcal{K}$ will be the smallest full subcategory of $\mathcal{S}$ such that

4.3.1. Every suspension of every object in σ lies in $\mathcal{K}$. That is, $\mathcal{K}$ contains all the complexes

$$\cdots \longrightarrow 0 \longrightarrow P_i \xrightarrow{s_i} Q_i \longrightarrow 0 \longrightarrow \cdots.$$

4.3.2. Given any chain map of objects in $\mathcal{K}$

$$\begin{array}{ccccccccc}
\cdots & \xrightarrow{\partial} & X^{i-1} & \xrightarrow{\partial} & X^{i} & \xrightarrow{\partial} & X^{i+1} & \xrightarrow{\partial} & \cdots \\
 & & \Big\downarrow{\scriptstyle f_{i-1}} & & \Big\downarrow{\scriptstyle f_i} & & \Big\downarrow{\scriptstyle f_{i+1}} & & \\
\cdots & \xrightarrow{\partial} & Y^{i-1} & \xrightarrow{\partial} & Y^{i} & \xrightarrow{\partial} & Y^{i+1} & \xrightarrow{\partial} & \cdots
\end{array}$$

then the mapping cone

$$\cdots \to X^i \oplus Y^{i-1} \xrightarrow{\begin{pmatrix} -\partial & 0 \\ f_i & \partial \end{pmatrix}} X^{i+1} \oplus Y^i \xrightarrow{\begin{pmatrix} -\partial & 0 \\ f_{i+1} & \partial \end{pmatrix}} X^{i+2} \oplus Y^{i+1} \to \cdots$$

also lies in $\mathcal{K}$.

As in Remark 4.2 we mean equality of chain complexes, not homotopy equivalence or homology isomorphism.

We note the obvious lemmas.

Lemma 4.4. Let $\tilde{\mathcal{K}}$ be the full subcategory of all objects in $\mathcal{S}$ isomorphic to objects in $\mathcal{K}$. That is, any object of $\mathcal{S}$ = D(A) isomorphic to a chain complex in $\mathcal{K}$ lies in $\tilde{\mathcal{K}}$. The subcategory $\tilde{\mathcal{K}} \subset \mathcal{S}$ is triangulated.

Proof. The point is that the objects of $\mathcal{K}$ are bounded chain complexes of projectives. Let $f: X \to Y$ be a morphism in D(A) between objects in $\mathcal{K}$. Because X is a bounded-above complex of projectives, there is a chain map representing the morphism. The mapping cone on this chain map completes $f: X \to Y$ to a triangle, and lies in $\mathcal{K}$. Up to isomorphism in $\mathcal{S}$ = D(A), all triangles on morphisms in $\mathcal{K}$ are contained in $\mathcal{K}$.

Lemma 4.5. The category $\mathcal{K}$ is contained in $\mathcal{R}^c$. Furthermore, every object in $\mathcal{R}^c$ is a direct

summand of an object isomorphic in D(A) to an object in $\mathcal{K}$.

Proof. The inclusion $\mathcal{K} \subset \mathcal{R}^c$ is easy. The category $\mathcal{R}^c$ contains σ and is closed under mapping cones, and $\mathcal{K}$ is the smallest such.

Next observe that, by [9, Lemma 2.2], the category $\mathcal{R}^c$ is the smallest thick subcategory of S containing σ, and hence $\mathcal{R}^c$ is the smallest thick subcategory containing the triangulated subcategory K of Lemma 4.4. Therefore every object of $\mathcal{R}^c$ is a direct summand of an object in $\mathcal{K}$.

The point of the exercise is that any object in $\mathcal{K}$ can be expressed as a mapping cone on a map of shorter objects. We need a definition, and then we are ready to state our main lemma.

Definition 4.6. The subcategories $\mathcal{K}$ [m,n] are defined as the intersection

$$\mathcal{K}[m,n] = \mathcal{K} \cap \mathcal{S}[m,n].$$

As in Definition 4.1, we allow m and n to be infinite.

Lemma 4.7. Suppose $n \in \mathbb{Z}$ is an integer. Then every object $Z \in \mathcal{K}$ can be expressed as a mapping cone on a chain map $Z_1 \to Z_2$, as below

$$\cdots \xrightarrow{\partial} Z_1^{n-1} \xrightarrow{\partial} Z_1^n \xrightarrow{\partial} Z_1^{n+1} \longrightarrow 0 \longrightarrow \cdots$$

$$\quad\quad\quad\quad \downarrow \quad\quad f_n\downarrow \quad\quad f_{n+1}\downarrow \quad\quad \downarrow$$

$$\cdots \longrightarrow 0 \longrightarrow Z_2^n \xrightarrow{\partial} Z_2^{n+1} \xrightarrow{\partial} Z_2^{n+2} \xrightarrow{\partial} \cdots$$

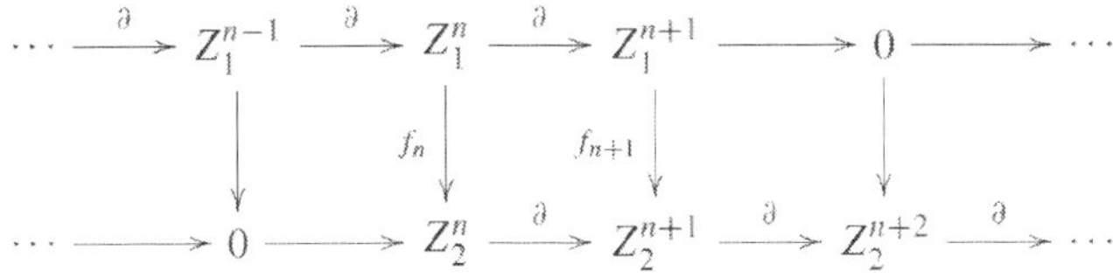

that is, $Z_1 \in \mathcal{K}(-\infty, n+1]$ and $Z_2 \in K[n, \infty)$.

Proof. Let B be the full subcategory of $\mathcal{K}$ containing the objects for which the assertion of the lemma holds. That is, an object $Z \in \mathcal{K}$ belongs to B if and only if, for every $n \in Z$, there exist $Z_1 \in \mathcal{K}(-\infty, n + 1]$ and $Z_2 \in \mathcal{K}[n, \infty)$ and a chain map $Z_1 \to Z_2$ so that Z is equal to the mapping cone. It suffices to prove that $B = \mathcal{K}$, for which we need only show that any suspension of an object of σ lies in B, and that mapping cones on maps in B lie in B.

Assume therefore that we are given a complex s below

$$\cdots \longrightarrow 0 \longrightarrow c^{\ell} \longrightarrow c^{\ell+1} \longrightarrow 0 \longrightarrow \cdots$$

which is some suspension of an object in σ. Choose any $n \in \mathbb{Z} \leq \ell$. If $n \leq \ell$, then $s \in \mathcal{K}[n,\infty)$, and s is the mapping cone of the chain map $0 \to s$. If $n \geq \ell + 1$, then $\Sigma^{-1}s \in \mathcal{K}(-\infty,n+1]$ and s is isomorphic to the mapping cone on the chain map $\Sigma^{-1}s \to 0$. Either way, $s \in \mathcal{B}$.

Next suppose we are given two object X and Y in $\mathcal{B}$, and a chain map $f : X \to Y$. Let Z be the mapping cone of f. We need to show that Z is in $\mathcal{B}$. For every integer $n \in Z$, we need to express Z as a mapping cone on a map of objects $Z_1 \to Z_2$, with $Z_1 \in \mathcal{K}(-\infty,n+1]$ and $Z_2 \in \mathcal{K}[n,\infty)$. Without loss of generality assume $n = 0$.

Because $X \in \mathcal{B}$ we may express it as a mapping cone on a map $X_1 \to X_2$, with $X_1 \in \mathcal{K}(-\infty, 2]$ and $X_2 \in \mathcal{K}[1,\infty)$. Because $Y \in \mathcal{B}$ we may express it as the mapping cone on a map $Y_1 \to Y_2$, with $Y_1 \in \mathcal{K}(-\infty, 1]$ and $Y_2 \in \mathcal{K}[0,\infty)$. We have a diagram, where the rows are short exact sequences of chain complexes

$$\begin{array}{ccccccccc} 0 & \longrightarrow & X_2 & \longrightarrow & X & \longrightarrow & \Sigma X_1 & \longrightarrow & 0 \\ & & & & \Big\downarrow{\scriptstyle f} & & & & \\ 0 & \longrightarrow & Y_2 & \longrightarrow & Y & \longrightarrow & \Sigma Y_1 & \longrightarrow & 0. \end{array}$$

The composite

$$\begin{array}{ccc} X_2 & \longrightarrow & X \\ & & \Big\downarrow{\scriptstyle f} \\ & & Y & \longrightarrow & \Sigma Y_1 \end{array}$$

is a chain map from $X_2 \in \mathcal{K}[1,\infty)$ to $\Sigma Y_1 \in \mathcal{K}(-\infty, 0]$, and therefore must vanish. It follows that we may complete to a commutative diagram of chain complexes

$$\begin{array}{ccccccccc} 0 & \longrightarrow & X_2 & \longrightarrow & X & \longrightarrow & \Sigma X_1 & \longrightarrow & 0 \\ & & \downarrow{\scriptstyle f_1} & & \downarrow{\scriptstyle f} & & \downarrow{\scriptstyle \Sigma f_2} & & \\ 0 & \longrightarrow & Y_2 & \longrightarrow & Y & \longrightarrow & \Sigma Y_1 & \longrightarrow & 0. \end{array}$$

Let Z_1 be the mapping cone on $f_1 : X_1 \to Y_1$, and let Z_2 be the mapping cone on $f_2 : X_2 \to Y_2$. Then $Z_1 \in \mathcal{K}(-\infty, 1]$ while $Z_2 \in \mathcal{K}[0,\infty)$. Furthermore Z, which is the mapping cone on $f : X \to Y$, can also be expressed as a mapping cone on a map $Z_1 \to Z_2$.

T INDUCES A *K*0-ISOMORPHISM

In this section we shall prove that the functor $T : \mathbf{T} \to \mathbf{U}$ of [10, Summary 5.4] induces an isomorphism $K_0(\mathbf{T}) \to K_0(\mathbf{U})$. Note that in higher K-theory there is a need to worry about models. But for any Waldhausen category $\mathbf{C}$, the Grothendieck group $K_0(\mathbf{C})$ is an invariant of the triangulated category $D(\mathbf{C})$. As long as we confine ourselves to K_0 computations we can quite safely work directly with the triangulated categories. This is what we will do in the current section.

Thus we must show that the functor of triangulated categories $T : \mathcal{T}^c \to D^{perf}(\sigma^{-1}A)$ induces an isomorphism in K_0. We shall do it through a sequence of lemmas. We resume the shorthands of Section 3. For every projective A-module we let $\sigma^{-1}P = \{\sigma + A\} \otimes_A P$, and we let $\mathcal{S} = D(A)$, $\mathcal{S}^c \cong D^{perf}(A)$ and $D = D^{perf}(\sigma^{-1}A)$. For the first time in this article we will need to occasionally mention unbounded derived categories. We remind the reader that the functor $\pi : D(A) \to \mathcal{T}$ has a right adjoint $G: \mathcal{T} \to D(A)$.

Lemma 5.1. Let n be an integer. Let $X \in \mathcal{S}^c$ be an object of $\mathcal{S}^{\le n}$, and let P be a finitely generated projective A-module. Then the functor $T : \mathcal{T}^c \to D$ gives a homomorphism

$$\mathcal{T}^c(\pi\Sigma^{-n}P, \pi X) \longrightarrow \mathcal{D}(T\pi\Sigma^{-n}P, T\pi X).$$

We assert that this map is an isomorphism.

Proof. By translation we may assume n = 0. We need to prove the map injective and surjective. Let us prove surjectivity first. Recall that $X \in \mathcal{S}^c \cap \mathcal{S}^{\le 0}$ is isomorphic to a chain complex of finitely generated projective A-modules

$$\to X^m \to X^{m+1} \to \cdots \to X^{-1} \to X^0 \to 0 \to 0 \to .$$

This makes $T\pi X$ the chain complex

$$\to \sigma^{-1}X^m \to \sigma^{-1}X^{m+1} \to \cdots \to \sigma^{-1}X^{-1} \to \sigma^{-1}X^0 \to 0 \to 0 \to .$$

Let P be a finitely generated projective A-module, concentrated in degree 0. Now the complex of $\sigma^{-1}A$-modules $T\pi P$ is a single projective module $\sigma^{-1}P$, concentrated in degree 0. Any map in the derived category, from the bounded above complex of projectives $\sigma^{-1}P = \{\sigma^{-1}A\} \otimes_A P$ to the complex $\{\sigma^{-1}A\}^L \otimes_A X$, can be represented by a chain-map. There is a map $\sigma^{-1}P \to \sigma^{-1}X_0$ inducing it. By Proposition 3.1, this comes from a map $\pi P \to \pi X^0$. But then the composite

$$\pi P \longrightarrow \pi X^0 \longrightarrow \pi X$$

gives a map $\pi P \to \pi X$ in $\mathcal{T}^c$, inducing $T\pi P \to T\pi X$.

This proved the surjectivity. For the injectivity, note that there is a short exact sequence of chain complexes

$$\begin{array}{ccccccccccccccc}
\to & 0 & \to & 0 & \to & \cdots & \to & 0 & \to & X^0 & \to & 0 & \to & 0 & \to \\
 & \downarrow & & \downarrow & & & & \downarrow & & \downarrow & & \downarrow & & \downarrow & \\
\to & X^m & \to & X^{m+1} & \to & \cdots & \to & X^{-1} & \to & X^0 & \to & 0 & \to & 0 & \to \\
 & \downarrow & & \downarrow & & & & \downarrow & & \downarrow & & \downarrow & & \downarrow & \\
\to & X^m & \to & X^{m+1} & \to & \cdots & \to & X^{-1} & \to & 0 & \to & 0 & \to & 0 & \to .
\end{array}$$

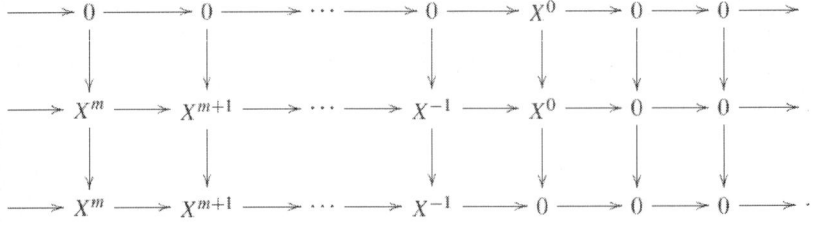

Write the corresponding triangle as

$$X^0 \longrightarrow X \longrightarrow Y \longrightarrow \Sigma X^0$$

The functor π takes this to the triangle

$$\pi X^0 \longrightarrow \pi X \longrightarrow \pi Y \longrightarrow \pi \Sigma X^0.$$

Let P be a finitely generated projective A-module, concentrated in degree 0. Suppose we are given a map $\pi P \to \pi X$. Composing to Y , we deduce a map

$$\pi P \longrightarrow \pi X \longrightarrow \pi Y.$$

By adjunction, this corresponds to a map

$$P \longrightarrow G\pi Y,$$

which must vanish. After all $Y \in \mathcal{S}^{\leq -1}$, and by [10, Lemma 6.4] it follows that $G\pi Y$ is also in $\mathcal{S}^{\leq -1}$. The map from a projective object P in degree 0 to the complex $G\pi Y \in \mathcal{S}^{\leq -1}$ has to vanish.

It follows that the map $\pi P \to \pi X$ must factor as

$$\pi P \longrightarrow \pi X^0 \longrightarrow \pi X.$$

Now assume that the map vanishes in $= D^c(\sigma^{-1}A)$. That is, the composite

$$\sigma^{-1} P \longrightarrow \sigma^{-1} X^0 \longrightarrow \{\sigma^{-1} A\}^L \otimes_A X$$

vanishes in $\mathcal{D}$. Then it must be null homotopic. The map $\sigma^{-1}P \to \sigma^{-1}X^0$ must factor as

$$\sigma^{-1} P \longrightarrow \sigma^{-1} X^{-1} \longrightarrow \sigma^{-1} X^0.$$

By Proposition 3.1, this tells us that the map $\pi P \to \pi X^0$ must factor as

$$\pi P \longrightarrow \pi X^{-1} \longrightarrow \pi X^0$$

and hence the map

$$\pi P \longrightarrow \pi X^{-1} \longrightarrow \pi X^{0} \longrightarrow \pi X$$

must vanish.

Lemma 5.2. Let n be an integer. Let Z be an object of $\mathcal{T}^{c}$, and suppose for all $r > n$, $H^{r}(T\ Z) = 0$. Then there is an object $X \in D^{perf}(A)$, that is a bounded complex of projective A-modules

$$\to 0 \to X^{m} \to X^{m+1} \to \cdots \to X^{\ell-1} \to X^{\ell} \to 0 \to$$

so that Z is a direct summand of an isomorph of πX, and $\ell \leq n$.

Proof. By suspending we may assume $n = 0$. Because every object of $\mathcal{T}^{c}$ is a direct summand of an isomorph of an object in $\mathcal{S}^{c}/\mathcal{R}^{c}$, we may certainly find an $X \in D^{perf}(A)$ and a $Z' \in \mathcal{T}^{c}$ with $\pi X \cong Z \oplus Z'$. What is not clear is that we may choose X to be a complex

$$\to 0 \to X^{m} \to X^{m+1} \to \cdots \to X^{\ell-1} \to X^{\ell} \to 0 \to$$

with $\ell \leq 0$. Assume therefore that $\ell > 0$, and we shall show that we may reduce ℓ by 1. We recall the short exact sequence of chain complexes

$$\begin{array}{ccccccccccccc}
\longrightarrow & 0 & \longrightarrow & 0 & \longrightarrow & 0 & \longrightarrow \cdots \longrightarrow & 0 & \longrightarrow & X^{\ell} & \longrightarrow & 0 & \longrightarrow \\
 & \downarrow & & \downarrow & & \downarrow & & \downarrow & & \downarrow & & \downarrow & \\
\longrightarrow & 0 & \longrightarrow & X^{m} & \longrightarrow & X^{m+1} & \longrightarrow \cdots \longrightarrow & X^{\ell-1} & \longrightarrow & X^{\ell} & \longrightarrow & 0 & \longrightarrow \\
 & \downarrow & & \downarrow & & \downarrow & & \downarrow & & \downarrow & & \downarrow & \\
\longrightarrow & 0 & \longrightarrow & X^{m} & \longrightarrow & X^{m+1} & \longrightarrow \cdots \longrightarrow & X^{\ell-1} & \longrightarrow & 0 & \longrightarrow & 0 & \longrightarrow
\end{array}$$

It gives a triangle which we write as

$$\Sigma^{-\ell} X^{\ell} \longrightarrow X \xrightarrow{a} Y \longrightarrow \Sigma^{-\ell+1} X^{\ell}.$$

We also have that Z is a direct summand of πX. That is, there are maps

$$\pi X \xrightarrow{b} Z \xrightarrow{c} \pi X$$

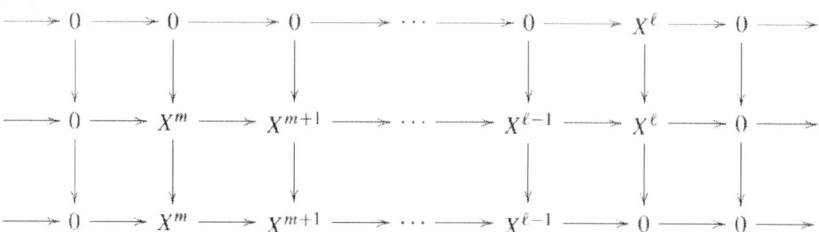

so that $bc = 1_Z$. Now we wish to consider the composite

$$\pi\Sigma^{-\ell}X^{\ell} \longrightarrow \pi X \xrightarrow{b} Z \xrightarrow{c} \pi X.$$

We know that X^{ℓ} is a finitely generated projective A-module, and $X \in S$ lies in $S^{\leq \ell}$. The conditions are as in Lemma 5.1. In order to prove that the composite vanishes, it suffices to prove that T of it vanishes, in $D = D^{perf}(\sigma^{-1}A)$.

But in D the map becomes the composite

$$T\pi\Sigma^{-\ell}X^{\ell} \longrightarrow T\pi X \longrightarrow TZ \longrightarrow T\pi X.$$

We assert that already the shorter composite, $T\pi\Sigma^{-\ell}X^{\ell} \to T\pi X \to TZ$ must vanish. After all, it is a map

$$T\pi\Sigma^{-\ell}X^{\ell} \longrightarrow TZ.$$

By hypothesis, TZ vanishes above degree 0. It is quasi-isomorphic to a complex of $\sigma^{-1}A$ modules in degree ≤ 0. And $T\pi\Sigma^{-\ell}X^{\ell} = \Sigma^{-\ell}\sigma - 1X^{\ell}$ is a single projective $\sigma^{-1}A$-module, concentrated in degree $\ell > 0$. Hence the vanishing. The composite

$$\pi\Sigma^{-\ell}X^{\ell} \longrightarrow \pi X \xrightarrow{b} Z \xrightarrow{c} \pi X$$

must therefore vanish. Since c is a split monomorphism, we deduce that the composite

$$\pi\Sigma^{-\ell}X^{\ell} \longrightarrow \pi X \xrightarrow{b} Z$$

also vanishes.

But now the triangle

$$\pi\Sigma^{-\ell}X^{\ell} \longrightarrow \pi X \xrightarrow{a} \pi Y \longrightarrow \pi\Sigma^{-\ell+1}X^{\ell}$$

tells us that the map b :$\pi X \to Z$ must factor as

$$\pi X \xrightarrow{a} \pi Y \xrightarrow{\beta} Z.$$

The composite

$$Z \xrightarrow{c} \pi X \xrightarrow{a} \pi Y \xrightarrow{\beta} Z$$

is the identity, and hence Z is a direct summand of an isomorph of πY, with Y the complex

$$\longrightarrow 0 \longrightarrow X^m \longrightarrow X^{m+1} \longrightarrow \cdots \longrightarrow X^{\ell-1} \longrightarrow 0 \longrightarrow 0 \longrightarrow \cdot$$

Lemma 5.3. Let n be an integer. Let Z be an object of T^c, and suppose for all r>n, $H^r(T\,Z) = 0$. Given any finitely generated projective A-module P , and any map

$$T\pi P = \sigma^{-1} P \xrightarrow{a} H^n(TZ),$$

there is a map in T^c

$$\pi \Sigma^{-n} P \xrightarrow{\mu} Z$$

so that $H^n(T\,\mu) = a$.

Proof. By translating we may assume n = 0. Let Z be an object of T^c and suppose that, for all r > 0, $H^r(T\,Z) = 0$. By Lemma 5.2 there exists a complex $X \in D^{perf}(A)$

$$\to 0 \to X^m \to X^{m+1} \to \cdots \to X^{-1} \to X^0 \to 0 \to$$

so that Z is a direct summand of an isomorph of πX. We have two maps

$$\pi X \xrightarrow{b} Z \xrightarrow{c} \pi X$$

so that $bc = 1_Z$. This gives us two maps

$$T\pi X \xrightarrow{Tb} TZ \xrightarrow{Tc} T\pi X$$

with (Tb)(Tc) = 1. Given any map

$$\sigma^{-1}P \xrightarrow{a} H^0(TZ),$$

we can form the composite

$$\sigma^{-1}P \xrightarrow{a} H^0(TZ) \xrightarrow{H^0(Tc)} H^0(T\pi X).$$

Of course, $T\pi X$ is just the chain complex

$$\cdots \to \sigma^{-1}X^{-1} \to \sigma^{-1}X^0 \to 0 \to$$

and any map from a projective $\sigma^{-1}P$ to $H^0(T\,\pi X)$ lifts to a map

$$\sigma^{-1}P \longrightarrow T\pi X.$$

By Lemma 5.1 the above map is $T\gamma$, for a (unique) map

$$\pi P \xrightarrow{\gamma} \pi X.$$

Now let μ be the composite

$$\pi P \xrightarrow{\gamma} \pi X \xrightarrow{b} Z.$$

Applying the functor $H^0 \circ T$, we compute $H^0(T\mu)$ to be the composite

$$\sigma^{-1}P \xrightarrow{a} H^0(TZ) \xrightarrow{H^0(Tc)} H^0(T\pi X) \xrightarrow{H^0(Tb)} H^0(TZ),$$

which is nothing other than the map a.

Lemma 5.4. For any finitely generated projective $\sigma^{-1}A$-module M, there is a canonically unique object $\tilde{M} \in T^c$ so that

$H^n(T\tilde{M}) = 0$ for $n \neq 0$.

$H^0(T\tilde{M}) = M$.

The functor $H^0(T\,-)$ is an equivalence of categories between the full subcategory of objects $\tilde{M} \in T^c$ and finitely generated projective $\sigma^{-1}A$-modules.

Proof. Let us first prove existence. Let M be a finitely generated projective $\sigma^{-1}A$-module. There exists a $\sigma^{-1}A$-module N, so that $M \oplus N \cong \{\sigma^{-1}A\}^r$. There is an idempotent $\{\sigma^{-1}A\}^r \to \{\sigma^{-1}A\}^r$ which is the map

$$M \oplus N \xrightarrow{1_M \oplus 0_N} M \oplus N.$$

Write this map as $1_M \oplus 0_N : T\pi A^r \to T\pi A^r$. By Proposition 3.1 there is a unique lifting $e : \pi A^r \to \pi A^r$. The uniqueness of the lifting allows us to easily show that $e^2 = e$. But idempotents split in T, by [2, Proposition 3.2 or Remark 3.3]. Define M by splitting the idempotent e.

Then $H^n(T\tilde{M})$ is computed by splitting the idempotent $H^n(T)$ on $H^n(\sigma^{-1}A^r)$; this gives us zero when $n \cong 0$, and M when $n = 0$. We have proved the existence of an $\tilde{M}$ satisfying 5.4.1 and 5.4.2.

Now suppose X is an object of T^c, and that

(i) $H^n(TX) = 0$ for $n \neq 0$,

(ii) $H^0(TX) = M$.

We wish to produce an isomorphism $\tilde{M} \to X$. In any case we have a map

$$\{\sigma^{-1}A\}^r \longrightarrow M = H^0(TX),$$

namely the projection to the direct summand. By Lemma 5.3 there is a map

$$\pi A^r \longrightarrow X$$

which induces the projection. We may form the composite

$$\tilde{M} \longrightarrow \pi A^r \longrightarrow X,$$

and it is very easy to check that the map

$$T\tilde{M} \longrightarrow TX$$

is a homology isomorphism, hence an isomorphism in $D^{perf}(\sigma^{-1}A)$. If we complete $\tilde{M} \to X$ to a triangle in T^c

$$\tilde{M} \longrightarrow X \longrightarrow Y \longrightarrow \Sigma\tilde{M},$$

then $TY = 0$. But by Proposition 3.4 it then follows that $Y = 0$, and $\tilde{M} \to X$ is an isomorphism. Finally it remains to check that $T^c(\tilde{M}, \tilde{N}) = \mathrm{Hom}_{\sigma^{-1}A}(M,N)$. By the construction of $\tilde{M}$ and $\tilde{N}$ as direct summands of πA^r and πA^s, this reduces to knowing that

$$\mathcal{T}^c(\pi A^r, \pi A^s) = \mathcal{D}(T\pi A^r, T\pi A^s)$$

is an isomorphism. But we know this from Proposition 3.1.

Theorem 5.5. The map T :$T^c \to D$ induces a K_0-isomorphism.

Proof. We define maps of categories

$$\mathcal{P}(\sigma^{-1}A) \xrightarrow{a} \mathcal{T}^c \xrightarrow{T} \mathcal{D}$$

with P(σ^{-1}A) the category of finitely generated projective σ^{-1}A-modules. The map T we already know. The map a takes a finitely generated projective σ^{-1}A-module M to a(M) = $\tilde{M}$. In K-theory, the composite

$$K_0(\sigma^{-1}A) \to K_0(\mathcal{T}^c) \to K_0(\mathcal{D})$$

is clearly an isomorphism. To prove that both maps are isomorphisms, it suffices to show that the map $K_0(a) : K_0(\sigma^{-1}A) \to K_0(T^c)$ is onto. This is what we shall do.

Let Z be an object of Tc. We want to show that its class [Z] ∈ $K_0(T^c)$ lies in the image of $K_0(a)$. We shall prove this by induction on the length of TZ. For the purpose of this proof, the length of TZ ∈ $D^{perf}(\sigma^{-1}A)$ is defined to be the smallest integer n for which there exists an integer m with

$$H^i(TZ) = 0 \quad \text{unless } m \leqslant i \leqslant m+n.$$

Suppose the length of TZ is zero. Replacing Z by a suspension, this means that $H^n(TZ) = 0$ unless n = 0. Since TZ ∈ $D^{perf}(\sigma^{-1}A)$, we have that $H^0(TZ)$ = M must be a finitely generated projective σ^{-1}A-module. By Lemma 5.4 we know that Z is (canonically) isomorphic to $\tilde{M}$. Thus Z is in the image of a.

Suppose now that we know the induction hypothesis. We are given n ≥0. We know that if Z is an object of T^c so that the length of TZ is ≤ n, then the class [Z] ∈ $K_0(T^c)$ lies in the image of $K_{0(}a)$. Let Z be a complex of length $n + 1 \geq 1$. Replacing Z by a suspension, this means that $H^r(TZ) = 0$ unless $-n - 1 \leq r \leq 0$. Now $H^0(TZ)$ is a finitely presented $\sigma^{-1}A$-module; we may choose a finitely generated free A-module F, and a surjection $\sigma^{-1}F \to H^0(TZ)$. By Lemma 5.3 there is a map

$$\pi F \longrightarrow Z$$

lifting this surjection. Form the triangle in T^c

$$\pi F \longrightarrow Z \longrightarrow Y \longrightarrow \Sigma \pi F.$$

It is easily computed that the length of TY is $\leq n$, so by induction [Y] lies in the image of $K_0(a) : K_0(\sigma^{-1}A) \to K_0(T^c)$. Clearly $[\pi F]=\widetilde{[\sigma^{-1}F]}$ lso lies in the image of $K_0(a)$, and the triangle tells us that $[Z]=[Y]+[\pi F]$.

T INDUCES A K1-ISOMORPHISM

In [10, Summary 5.4] we constructed a diagram of Waldhausen models

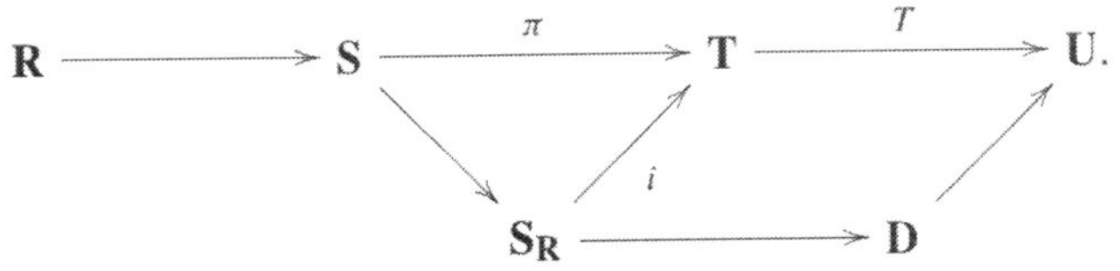
R
S
π
T
T
U.
i
S_R
D

On applying the K-theory functor, we deduce a diagram

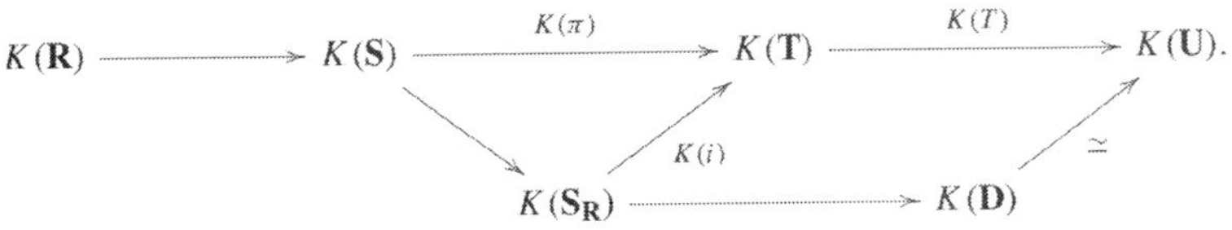
K(R)
K(S)
K(π)
K(T)
K(T)
K(U).
K(i)
≃
K(S_R)
K(D)

What we want to show is that $K_1(T) \simeq K_1(D)$. As the diagram indicates, we already know that the natural map $K(D) \to K(U)$ is a homotopy equivalence. I want to remind the reader that Grayson's cofinality theorem [10, Theorem 2.4] tells us that the map $K_1(i):K_1(S_R) \to K_1(T)$ is also an isomorphism. It is only at the level of K_0 that the map K(i) fails to be an isomorphism.

The diagrams above therefore reduce us to showing that the natural map $K_1(S_R) \to K_1(D)$ is an isomorphism. The advantage is that we do not have to work with the models T and U, which involve unbounded complexes.

In the proof of Theorem 5.5 we introduced a functor $P(\sigma^{-1}A) \to T^c$. We do not know a Waldhausen model for this map. But we shall now show that there is an induced map on K_1. The group $K_1(\sigma^{-1}A)$ is generated by determinants of automorphisms of free (or projective) modules. This means: Given any projective A-module P , and an automorphism $\phi : \sigma^{-1}P \to \sigma^{-1}P$, the determinant of ϕ is an element of $K_1(\sigma^{-1}A)$. The collection of all determinants of all ϕ's generates $K_1(\sigma^{-1}A)$. We want to produce a map $K_1(\sigma^{-1}A) \to K_1(S_R)$; to0 define the map, it suffices to say what it does on all ϕ's as above.

To define what the map does to ϕ, let us remind ourselves that the zero-space of the spectrum $K(S_R)$ has a Gillet–Grayson model (see [6]), which we denote $GG(S_R)$. That is, there is a homotopy equivalence

$$GG(\mathbf{S_R}) \simeq \Omega^{\infty} K(\mathbf{S_R}).$$

The space $GG(S_R)$ is an H-space, and hence

$$K_1(\mathbf{S_R}) = \pi_1 K(\mathbf{S_R}) = \pi_1 GG(\mathbf{S_R}) = H_1 GG(\mathbf{S_R}).$$

Starting with an automorphism $\phi : \sigma^{-1}P \to \sigma^{-1}P$, we need to produce a class in the first homology group $H^1GG(S_R)$.

We note that, by Proposition 3.1, $\phi : \sigma^{-1}P \to \sigma^{-1}P$ corresponds to a unique automorphism

$$\varphi : \pi P \to \pi P.$$

This is an automorphism defined in S^c/R^c, and S_R is a Waldhausen model for S^c/R^c. It follows that there exist weak equivalences $a: Q \to P$ and $b: Q \to P$, with $\varphi = ab^{-1}$. But then $(P, 0)$ and $(Q, 0)$ are 0-cells in the Gillet–Grayson model $GG(S_R)$. The weak equivalences $(a, 0):(Q, 0) \to (P, 0)$ and $(b, 0):(Q, 0) \to (P, 0)$ are 1-cells. The difference, which we denote $[a]-[b]$, is a cycle. It is an element in $H_1(GG(S_R)) = \pi_1(GG(S_R)) = K_1(S_R)$. We leave it to the reader to check that the map sending ϕ to $[a]-[b]$ extends to a well-defined homomorphism $\psi : K_1(\sigma^{-1}A) \to K_1(S_R)$.

Recall that $D = C^{perf}(\sigma^{-1}A)$, and that the map $S_R \to D$ is just tensor product with $\sigma^{-1}A$. The composite

$$K_1(\sigma^{-1}A) \xrightarrow{\psi} K_1(\mathbf{S_R}) \xrightarrow{\theta} K_1(\mathbf{D})$$

is easily seen to be the K_1 part of the natural map $K(\sigma^{-1}A) \to K(D) = K(C^{perf}(\sigma^{-1}A))$, and by Gillet's [5, 6.2] we know it to be an isomorphism. To prove that both ψ and θ are isomorphisms it suffices to check that $\psi : K_1(\sigma^{-1}A) \to K_1(S_R)$ is epi. We have a localisation exact sequence

$$K_1(\mathbf{S}) \longrightarrow K_1(\mathbf{S_R}) \longrightarrow K_0(\mathbf{R}) \longrightarrow K_0(\mathbf{S}).$$

Note that $K(S) = K(C^{perf}(A)) = K(A)$. The composite $K_1(A) \to K_1(\sigma^{-1}A) \to K_1(S_R)$ is easily computed to agree with the natural $K_1(A) = K_1(S) \to K_1(S_R)$; we deduce a commutative diagram where the bottom row is exact

$$\begin{array}{ccccccc} K_1(A) & \longrightarrow & K_1(\sigma^{-1}A) & & & & \\ \downarrow{\scriptstyle 1} & & \downarrow{\scriptstyle \psi} & & & & \\ K_1(A) & \longrightarrow & K_1(\mathbf{S_R}) & \longrightarrow & K_0(\mathbf{R}) & \longrightarrow & K_0(A). \end{array}$$

To prove ψ epi, it suffices to show that the composite

$$\begin{array}{ccc} K_1(\sigma^{-1}A) & & \\ \downarrow \psi & & \\ K_1(\mathbf{S_R}) & \longrightarrow & K_0(\mathbf{R}) \end{array}$$

surjects to the kernel of $K_0(R) \to K_0(A)$. But the composite is easy to compute. Take an automorphism $\phi : \sigma^{-1}A \otimes P \to \sigma^{-1}A \otimes P$ as above, which corresponds as above to an automorphism

$$\varphi : \pi P \to \pi P.$$

Choose weak equivalences $a : Q \to P$ and $b : Q \to P$, with $\varphi = ab^{-1}$. Then ϕ gets sent to $[A]-[B]$, where

$$A: \quad \cdots \to 0 \to Q \xrightarrow{a} P \to 0 \to \cdots,$$

$$B: \quad \cdots \to 0 \to Q \xrightarrow{b} P \to 0 \to \cdots.$$

It will therefore suffice to show that every element in the kernel of the map $K_0(R) \to K_0(A)$ can be expressed as a difference $[A]-[B]$ as above. Note that this is a statement about K_0, and K_0 is an invariant of the derived category, independent of choices of models. From now on we can forget all about models. We shall prove something a little stronger than we need.

Theorem 6.1. Every element in $K_0(R) = K_0(R^c)$ is a linear combination of complexes of length ≤ 1. That is, it may be written $\Sigma \pm [A_i]$, with $A_i \in R^c$ being complexes of finitely generated projective A-modules of the form

$$\cdots \to 0 \to X \to Y \to 0 \to \cdots.$$

Before the proof of Theorem 6.1, let us state the main corollary.

Corollary 6.2. Every object in the kernel of the map $K_0(R^c) \to K_0(S^c) = K_0(A)$ is of the form [A]−[B], where A is a complex

$$\cdots \to 0 \to Q \xrightarrow{a} P \to 0 \to \cdots$$

and B is a complex

$$\cdots \to 0 \to Q \xrightarrow{b} P \to 0 \to \cdots.$$

By the discussion preceding Theorem 6.1 this means that the map $S_R \to D$ induces a K_1- isomorphism.

Proof that Corollary 6.2 follows from Theorem 6.1. Suppose we have an element of the kernel of the map $K_0(R^c) \to K_0(S^c) = K_0(A)$. By Theorem 6.1, just by virtue of being an element of $K_0(R^c)$, it has an expression as $\Sigma \pm[A_i]$, with $A_i \in R^c$ being complexes of finitely generated projective A-modules of the form

$$\cdots \to 0 \to X_i \to Y_i \to 0 \to \cdots.$$

Recalling that $[\Sigma A_i] = -[A_i]$, up to changing signs in the sum we may assume that all the X_i are in degree −1, all the Y_i in degree 0. Collecting together all the terms of equal sign, we may rewrite the sum as

$$\left[\bigoplus A_i\right] - \left[\bigoplus B_j\right].$$

That is, we have an element [A]−[B] in the kernel of $K_0(R^c) \to K_0(S^c) = K_0(A)$, where A,B are complexes of the form

$$A: \quad \cdots \to 0 \to A^{-1} \xrightarrow{a} A^0 \to 0 \to \cdots,$$

$$B: \quad \cdots \to 0 \to B^{-1} \xrightarrow{b} B^0 \to 0 \to \cdots.$$

The fact that [A]−[B] lies in the kernel of the map $K_0(R^c) \to K_0(A)$ tells us that, in $K_0(A)$, there is an identity

$$[A^{-1}]+[B^0]=[B^{-1}]+[A^0].$$

This in turn says that there is a projective A-module X, and an isomorphism

$$A^{-1} \oplus B^0 \oplus X \cong B^{-1} \oplus A^0 \oplus X.$$

The object $[A] \in R^c$ is isomorphic to the complex

$$\cdots \longrightarrow 0 \longrightarrow A^{-1} \oplus B^0 \oplus X \xrightarrow{a\oplus 1_{B^0}\oplus 1_X} A^0 \oplus B^0 \oplus X \longrightarrow 0 \longrightarrow \cdots$$

while the object $[B] \in R^c$ is isomorphic to the complex

$$\cdots \longrightarrow 0 \longrightarrow B^{-1} \oplus A^0 \oplus X \xrightarrow{b\oplus 1_{A^0}\oplus 1_X} B^0 \oplus A^0 \oplus X \longrightarrow 0 \longrightarrow \cdots.$$

Put $Q = A^{-1} \oplus B^0 \oplus X \cong B^{-1} \oplus A^0 \oplus X$, and $P = A^0 \oplus B^0 \oplus X$. Then A is isomorphic in R^c to a complex

$$\cdots \to 0 \to Q \xrightarrow{\alpha} P \to 0 \to \cdots$$

and B is isomorphic in Rc to a complex

$$\cdots \to 0 \to Q \xrightarrow{\beta} P \to 0 \to \cdots$$

as required.

Proof of Theorem 6.1. It remains to prove Theorem 6.1. Let X be an object of R^c. We need to show that the class [X] in $K_0(R^c)$ can be written as a linear combination of classes of objects of length ≤ 1.

Because $X \in R^c \subset S^c$, we have that X is isomorphic to a bounded complex of finitely generated projective A-modules. Suspending suitably and replacing by an isomorph, we may assume X has the form

$$\to 0 \to X^{-m} \to X^{-m+1} \to \cdots \to X^{-1} \to X^0 \to 0 \to .$$

If $m \leq 1$ we are done; the complex has length ≤ 1. The proof is by induction on m. Assume we are given an integer $n \geq 1$. Assume further that, for every $X \in R^c$ of length $m \leq n$, [X] is equal in $K_0(R^c)$ to a linear combination of complexes of length ≤ 1. Take a complex X as above, with $m = n + 1 \geq 2$. We need to show that it can also be expressed as a linear combination of complexes of length ≤ 1.

By Lemma 4.5 we have that every object of R^c is a direct summand of an isomorph of an object in K, with K as in Definition 4.3. Choose a chain complex $Y \in K$ and maps $X \to Y \to X$ composing to the identity on X. Clearly the map $H^0(Y) \to H^0(X)$ must be surjective.

By Lemma 4.7 there exist exists a triangle

$$U \longrightarrow Y \longrightarrow V \longrightarrow \Sigma U$$

with $U \in K[1,\infty)$ and $V \in K(-\infty, 1]$. The composite

$$U \longrightarrow Y \longrightarrow X$$

is a map from $U \in K[1,\infty)$ to $X \in S{\leq}0$, which must vanish. It follow that $Y \to X$ factors as $Y \to V \to X$. And since $H^0(Y) \to H^0(X)$ is epi and factors through $H^0(V)$, we deduce that $H^0(V) \to H^0(X)$ must be epi. Replacing Y by V we may assume $Y \in K(-\infty, 1]$.

Next we apply Lemma 4.7 again, this time to deduce that $Y \in K(-\infty, 1]$ can be expressed as the mapping cone on a map $U \to V$, with $U \in K(-\infty, 0]$ and $V \in K[-1, 1]$. There is a triangle

$$U \longrightarrow V \longrightarrow Y \longrightarrow \Sigma U,$$

hence an exact sequence

$$H^0(V) \longrightarrow H^0(Y) \longrightarrow H^1(U) = 0.$$

The map $H^0(V) \to H^0(Y) \to H^0(X)$ is the composite of two epis, hence is epi. Replacing $Y \to X$ by the composite $V \to Y \to X$, we may assume $Y \in K[-1, 1]$.

The last time we apply Lemma 4.7 is to express Y as the mapping cone of a map $U \to Z$, with $U \in K[0, 1]$ and $Z \in K[0, 1]$. The only observations we wish to make is that $H^1(Z) \to H1(Y)$ is epi, and that in $K_0(R^c)$, [Z] and [Y]=[Z]−[U] are both linear combinations of objects in R^c of length ≤ 1. Let us summarise: we have constructed maps $Z \to Y \to X$, with

(i) $Z \in K[0, 1]$, $Y \in K[-1, 1]$,

(ii) $H^0(Y) \to H^0(X)$ epi,

(iii) $H^1(Z) \to H^1(Y)$ epi,

(iv) both [Y] and [Z] are linear combinations of objects in R^c of length ≤ 1.

Form the mapping cone on the map $Y \to X$ to obtain a triangle

$$Y \longrightarrow X \longrightarrow X' \longrightarrow \Sigma Y.$$

Since $Y \in K[-1, 1]$ while $X \in Rc$ is supported on the interval $[-m, 0]$ with $m \geq 2$, the mapping cone X′ is an object of R^c supported in $[-m, 0]$. The long exact sequence in homology gives

$$H^0(Y) \longrightarrow H^0(X) \longrightarrow H^0(X') \longrightarrow H^1(Y) \longrightarrow H^1(X).$$

We have $H^1(X) = 0$, while $H^0(Y) \to H^0(X)$ is an epimorphism. Hence $H^0(X') = H^1(Y)$. But we know that the map $H^1(Z) \to H^1(Y)$ is an epimorphism, by (iii). And Z is a complex of the form

$$\cdots \to 0 \to Z^0 \to Z^1 \to 0 \to \cdots$$

that is a complex of length ≤ 1. It follows that we can extend the epimorphism $\beta : H^1(Z) \to H^0(X')$ to a map from the presentation Z; there is a map $\Sigma Z \to X'$, inducing β in H^0. We may form the mapping cone, obtaining a triangle

$$\Sigma Z \longrightarrow X' \longrightarrow X'' \longrightarrow \Sigma^2 Z.$$

Since $\Sigma Z \in K[-1, 0]$ and X' is supported on $[-m, 0]$ with $m \geq 2$, we conclude that X'' is supported on $[-m, 0]$. But now the long exact homology sequence

$$H^0(\Sigma Z) \xrightarrow{\alpha} H^0(X') \longrightarrow H^0(X'') \longrightarrow H^1(\Sigma Z)$$

has α surjective, while $H^1(\Sigma Z) = 0$. We conclude that $H^0(X'') = 0$. The complex X'' may be written as

$$\to 0 \to \widetilde{X}^{-m} \to \widetilde{X}^{-m+1} \to \cdots \to \widetilde{X}^{-1} \to \widetilde{X}^0 \to 0 \to$$

and since $H^0(X'') = 0$ the map $\tilde{X}^{-1} \to \tilde{X}^0$ is an epimorphism. Since X^0 is projective, this epimorphism must be split. The complex X'' is homotopy equivalent to a complex

$$\to 0 \to \widetilde{X}^{-m} \to \widetilde{X}^{-m+1} \to \cdots \to \widetilde{Y}^{-1} \to 0 \to 0 \to$$

supported in the interval $[-m,-1]$. By induction, its class in $K_0(R^c)$ is a linear combination of complexes of length ≤ 1.

But now the triangles above give the identities

$$[X] = [X'] + [Y], \qquad [X'] = [X''] - [Z]$$

and hence $[X]=[X'']+[Y]-[Z]$, and all the terms on the right may be expressed as linear combinations of complexes of length ≤ 1.

ACKNOWLEDGMENT

The author would like to acknowledge the contribution of Andrew Ranicki. Until very recently Ranicki was a coauthor of the article. For reasons that were never entirely clear to the me, Ranicki chose to withdraw as an author. The author was partly supported by Australian Research Council grant DP0343239.

REFERENCES

1. Paul Balmer, Marco Schlichting, Idempotent completion of triangulated categories, J. Algebra 236 (2) (2001) 819– 834.
2. Marcel Bökstedt, Amnon Neeman, Homotopy limits in triangulated categories, Compos. Math. 86 (1993) 209–234.
3. P.M. Cohn, Free ideal rings, J. Algebra 1 (1964) 47–69.
4. William G. Dwyer, Noncommutative localization in homotopy theory, preprint http://www.nd.edu/~wgd/.
5. Henri Gillet, Riemann–Roch theorems for higher algebraic K-theory, Adv. Math. 40 (3) (1981) 203–289.
6. Henri Gillet, Daniel R. Grayson, The loop space of the Q-construction, Illinois J. Math. 31 (4) (1987) 574–597.
7. Daniel R. Grayson, Higher algebraic K-theory. II (after Daniel Quillen), in: Algebraic K-Theory, Proc. Conf., Northwestern Univ., Evanston, IL, 1976, in: Lecture Notes in Math., vol. 551, Springer, Berlin, 1976, pp. 217–240.
8. Henning Krause, Cohomological quotients and smashing localizations, http://wwwmath.upb.de/~hkrause/publications.html.
9. Amnon Neeman, The connection between the K-theory localisation theorem of Thomason, Trobaugh and Yao, and the smashing subcategories of Bousfield and Ravenel, Ann. Sci. École Norm. Sup. 25 (1992) 547–566.
10. Amnon Neeman, Andrew Ranicki, Noncommutative localisation in algebraic K-theory I, Geom. Topol. 8 (2004) 1385–1425.

11. Amnon Neeman, Andrew Ranicki, Noncommutative localisation in algebraic L-theory, http://wwwmaths.anu. edu.au/~neeman/preprints.html.
12. Amnon Neeman, Andrew Ranicki, Aidan Schofield, Representations of algebras as universal localizations, Math. Proc. Cambridge Philos. Soc. 136 (1) (2004) 105–117.
13. Aidan H. Schofield, Representation of Rings over Skew Fields, Cambridge Univ. Press, Cambridge, 1985.
14. Charles A. Weibel, Dongyuan Yao, Localization for the K-theory of noncommutative rings, in: Algebraic K-Theory, Commutative Algebra, and Algebraic Geometry, Santa Margherita Ligure, 1989, Amer. Math. Soc., Providence, RI, 1992, pp. 219–230.

CITATION

1. Amnon Neeman, Noncommutative localisation in algebraic K-theory II, Advances in Mathematics, Volume 213, Issue 2, 20 August 2007, Pages 785-819, ISSN 0001-8708, http://dx.doi.org/10.1016/j.aim.2007.01.010.

Lie Algebras and Lie Groups over Noncommutative Rings

Arkady Berenstein[a] *and*
Vladimir Retakh[b]
[a]Department of Mathematics, University of Oregon, Eugene, OR 97403, USA
[b]Department of Mathematics, Rutgers University, Piscataway, NJ 08854, USA

7

ABSTRACT

The aim of this paper is to introduce and study Lie algebras and Lie groups over noncommutative rings. For any Lie algebra g sitting inside an associative algebra A and any associative algebra F we introduce and study the algebra (g,A)(F), which is the Lie subalgebra of $F \otimes A$ generated by $F \otimes g$. In many examples A is the universal enveloping algebra of g. Our description of the algebra (g,A)(F) has a striking resemblance to the commutator expansions of F used by M. Kapranov in his approach to noncommutative geometry. To each algebra (g,A)(F) we associate a "noncommutative algebraic" group which naturally acts on (g,A)(F) by conjugations and conclude the paper with some examples of such groups.

INTRODUCTION

The aim of this paper is to introduce and study algebraic groups and Lie algebras over noncommutative rings.

Our approach is motivated by the following considerations. A naive definition of a Lie algebra as a bimodule over a noncommutative associative algebra $\mathcal{F}$ (over a field $\mathbb{k}$) does not bring any interesting examples beyond Lie algebra $gl_n(\mathcal{F})$. Even the special Lie algebra $sl_n(\mathcal{F}) = [gl_n(\mathcal{F}), gl_n(\mathcal{F})]$ is not an $\mathcal{F}$-bimodule. Similarly, the special linear group $SL_n(\mathcal{F})$ is not defined by equations but rather by congruences given by the Dieudonné determinant (see [1]). This is why the "straightforward" approach to classical groups over rings started by J. Dieudonné in [4] and continued by O.T. O'Meara and others (see [6]) does not lead to new algebraic groups. Also, unlike in the commutative case, these methods do not employ rich structural theory of Lie algebras.

As a starting point, we observe that the Lie algebra $sl_n(\mathcal{F})$ where $\mathcal{F}$ is an associative algebra over a field $\Bbbk$ (of characteristic 0) is the Lie subalgebra of $M_n(\mathcal{F}) = F \otimes M_n(k)$ generated by $F \otimes sl_n(\Bbbk)$ (all tensor products in the paper are taken over k unless specified otherwise). This motivates us to consider, for any Lie subalgebra $\mathfrak{g}$ of an associative algebra A, the Lie subalgebra

$$(\mathfrak{g}, A)(\mathcal{F}) \subset \mathcal{F} \otimes A \tag{0.1}$$

Generated by $F \otimes \mathfrak{g}$.

If F is commutative, then $\mathcal{F} \otimes \mathfrak{g}$ is already a Lie algebra, and $(\mathfrak{g},A)(\mathcal{F}) = \mathcal{F} \otimes \mathfrak{g}$. However, if $\mathcal{F}$ is noncommutative, this equality does not hold. Our first main result (Theorem 4.3) is a formula expressing $(\mathfrak{g},A)(\mathcal{F})$ in terms of powers $\mathfrak{g}^n = \text{Span}\{g_1 g_2 \dots g_n : g_1,\dots,g_n \in \mathfrak{g}\}$ in A for all perfect pairs $(\mathfrak{g},A)$ in the sense of Definition 4.1. The class of perfect pairs is large enough—it includes all semisimple and Kac–Moody Lie algebras $\mathfrak{g}$.

More precisely, Theorem 4.3 states that for any perfect pair $(\mathfrak{g},A)$ and any associative algebra $\mathcal{F}$ we have

$$(\mathfrak{g}, A)(\mathcal{F}) = \mathcal{F} \otimes \mathfrak{g} + \sum_{n \geqslant 1} I_n \otimes [\mathfrak{g}, \mathfrak{g}^{n+1}] + [\mathcal{F}, I_{n-1}] \otimes \mathfrak{g}^{n+1}, \tag{0.2}$$

where $I_0 \supset I_1 \subset I_2 \supset \cdot$ is the descending filtration of two-sided ideals of $\mathcal{F}$ defined inductively as $I_0 = \mathcal{F}$, $I_n+1 = \mathcal{F}\ [\mathcal{F}, I_n]+[\mathcal{F}, I_n]$. It is remarkable that this filtration emerged in works by M. Kapranov in [7] and then by M. Kontsevich and A. Rosenberg in [8] as an important tool in noncommutative geometry.

We also get a compact formula (Theorem 4.9) for $(sl_2(\Bbbk),A)(\mathcal{F})$ which, hopefully, has physical implications.

For unital F we define the "noncommutative" current group or, in short, the N -current group $G_{g',A}(F)$ to be the set of all invertible elements X $\in$ F $\otimes$ A such that $X \cdot (g,A)(F) \cdot X^{-1} = (g,A)(F)$. This is our generalization of $GL_n(F)$. In fact, if $g = sl_n(k) \subset M_n(k) = A$ then $G_{g'}A(F) = GL_n()$.

However, for other compatible pairs the structure of $G_{g'}A(F)$ is rather nontrivial even for classical Lie algebras $g = o_n(k)$ and $g = sp_n(k)$ and A $= M_n(k)$. To demonstrate this, we explicitly compute the "Cartan subgroup" of $G_{g,A}(F)$ (Proposition 5.11) as follows. For the above classical compatible pair (g,A) an invertible diagonal matrix $D = diag(f_1,\dots,f_n) \in M_n(F)$ belongs to the $G_{g,A}(F)$ if and only if\

$$f_i f_{n-i+1} - f_1 f_n \in I_1 = \mathcal{F}[\mathcal{F}, \mathcal{F}]$$

for i = 1,...,n

Our computation of the "Cartan subgroup" for $g = sl_2()$ and $A = M_n()$ is dramatically harder and constitutes the second main result of this paper (Theorem 5.12). More precisely, let $g = sl_2() \subset End_{\mathbb{K}}(V) = M_n(k) = A$, where $V = S^{n-1}(\mathbb{K}^2)$ is the simple n-dimensional $sl^2(\mathbb{K})$- module. Then an invertible diagonal matrix $diag(f_1,f_2,\dots,f_n)$ belongs to $G_{g,A}(F)$ if and only if

$$\sum_{i=0}^{m}(-1)^i\binom{m}{i} f_{i+1} f_{i+2}^{-1} \in I_m \tag{0.3}$$

For m = 1, 2,...,n − 2.

We can also apply our functorial generalization of $GL_n(F)$ to K-theory (however, we postpone all computations for concrete rings F until a separate paper). We propose to generalize the fundamental inclusion which plays a pivotal role in the algebraic K-theory

$$E_n(\mathcal{F}) \subset GL_n(\mathcal{F}) \tag{0.4}$$

Where $E_n(F)$ is the subgroup generated by matrices $1 + {}_tE_{ij}$, $t \in F$, $I \neq j$. Here the E_{ij} are elementary matrices. It is well known and widely used that $E_n(F)$ is normal in $GL_n(F)$ for $n \geq 3$ (and for certain algebras F when $n = 2$).

It turns out that both groups $E_n(F)$ and $GL_n(F)$ are completely determined by the compatible pair (g,A), where $g := sl_n(k) \subset M_n(k) = A$. Then $sl_n(F) = (g,A)(F)$, $GL_n(F) = G_{g,A}(F)$ and the group $E_n(F)$ is generated by all elements $g \in G_{g,A}(F)$ of the form $g = 1+t \otimes E$ where $E \in g$ is a nilpotent in A, $t \in F$.

Motivated by this observation, we propose to generalize the inclusion (0.4) to all of our N - current groups as follows. We define (Section 5.1) the group $E_{g,A}(F)$ to be the sub-group of $G_{g,A}(F)$ generated by all elements of the form $1 + {}_t\otimes{}_s$ where $t \in F$ and $s \in g$ is a nilpotent in A. Clearly, $E_{g,A}(F)$ is a normal subgroup of $G_{g,A}(F)$ which plays the role of $E_n(F)$ in $GL_n(F)$. This defines a generalized K_{1-}functor

$$\mathcal{F} \mapsto K_1^{\mathfrak{g},A}(\mathcal{F}) = G_{\mathfrak{g},A}(\mathcal{F})/E_{\mathfrak{g},A}(\mathcal{F}).$$

In Section 5.1 we also construct other generalized K_1-functors in which we replace $E_{g,A}(F)$ by $ES_{,g,A}(F)$, the subgroup generated by 1+S, where S is a $G_{g,A}(F)$-invariant subset of nilpotents in $F \otimes A$. However, computation of the generalized K_{1-}functors is beyond the scope of the present paper and will be performed in a separate publication.

The paper is organized as follows.

- Section 1 contains some preliminary results on ideals in associative algebras F generated by kth commutator spaces of F. Several key results are based on the Jacobi–Leibniz type identity (1.5).

- In Section 2 we introduce N -Lie algebras and their important subclass: N -current Lie algebras (g,A)(F) over Lie algebras g. As our first examples, we describe algebras (g,A)(F) for all classical Lie algebras.
- Section 3 contains upper bounds for N -current Lie algebras.
- Section 4 contains our main result for Lie algebras (g,A)(F): upper bounds for algebras (g,A)(F) coincide with them for a large class of compatible pairs (g,A) including all such pairs for semisimple Lie algebras g.
- In Section 5 we introduce affine N -groups and N -current groups, there relation with N - Lie algebras and N -current Lie algebras, important classes of their normal subgroups similar to subgroups $E_n(F)$ and the corresponding K_1-functors. We also consider useful examples of N -subgroups and their "Cartan subgroups" attached to the standard representations of classical Lie algebras g and to various representations of $g = sl_2(\mathbb{K})$. Our description of these subgroups is based on a new class of algebraic identities for noncommutative difference derivatives (Lemma 5.25) which are of interest by themselves.

The present paper continues our study of algebraic groups over noncommutative rings and their representations started in [2]. Part of our results was published in [3]. In the next paper we will focus on "reductive groups over noncommutative rings," their geometric structure and representations.

Throughout the paper, Alg will denote the category which objects are associative algebras (not necessarily with 1) over a field $\mathbb{K}$ of characteristic zero and morphisms are algebra homomorphisms; and N will stand for a sub-category of Alg. Also Alg_1 will denote that sub-category of Alg which objects are unital $\mathbb{K}$-algebras over $\mathbb{K}$ and arrows are homomorphisms of unital algebras.

COMMUTATOR EXPANSIONS AND IDENTITIES

Given an object F ∈ Alg, for k≥ 0 define the kth commutator space $F^{(k)}$ of F recursively as $F^{(0)} = F$, $F^{(1)} = F = [F,F]$, $F^{(2)} = F = [F,F\,]$, … , $F^{(k)} = [F,F^{(k-1)}\,]$, … , where for any subsets S_1,S_2 of F the notation $[S_1,S_2]$ stands for the linear span of all commutators $[a,b] = ab - ba$, $a \in S_1$, $b \in S_2$. In a similar fashion, for each subset $S \subset F$ define the subspaces $S^{(k)} \subset F^{(k)}$ by $S^{(0)} = span^{(S)}$, $S^{(k)} = [S,S^{(k-1)}\,]$ for $k > 0$; and the subspace $S^{(\bullet)}$ in F by

$$S^{(\bullet)} = \sum_{k \geqslant 0} S^{(k)}. \tag{1.1}$$

The following result is obvious

Lemma 1.1. For any $S \subset F$ the subspace $S^{(\bullet)}$ is the Lie subalgebra of F generated by S.

Following [7] and [8], define the subspaces I_k^ℓ (F) by:

$$I_k^\ell(\mathcal{F}) = \sum_\lambda \mathcal{F}^{(\lambda_1)}\mathcal{F}^{(\lambda_2)} \ldots \mathcal{F}^{(\lambda_\ell)}$$

Where the summation goes over all $\lambda = (\lambda_1,\lambda_2,\ldots,\lambda_\ell) \in (\mathbb{Z} \geq 0)^\ell$ such that $\sum_{\ell=1} I_k^\ell\ \lambda_i = k$. Denote also

$$I_k^{\leqslant \ell}(\mathcal{F}) := \sum_{1 \leqslant \ell' \leqslant \ell} I_k^{\ell'}(\mathcal{F}), \qquad I_k(\mathcal{F}) := I_k^{<\infty} = \sum_{\ell \geqslant 1} I_k^\ell(\mathcal{F}). \tag{1.2}$$

Clearly, FI_k^ℓ (F), I_k^ℓ (F)F $\subset \sum_{\ell=1} I_k^{\ell+1}$ k (F). Therefore, I_k(F) is a two-sided ideal in F. Taking into account that $F^{(k)}F + F^{(k)} = FF^{(k)} + F^{(k)}$ for all k, it is easy to see that $I_k(F) = I_k^k$ (F) + FI_k^k (F) for all k.

Lemma 1.2. For each k, $\ell \geq 1$ one has:

a. $I_k^\ell(\mathcal{F}) \subset I_{k-1}^\ell(\mathcal{F}),\ I_k^{\leqslant \ell}(\mathcal{F}) \subset I_{k-1}^{\leqslant \ell}(\mathcal{F}).$

b. $[\mathcal{F}, I_{k-1}^\ell(\mathcal{F})] \subset I_k^\ell(\mathcal{F}),\ [\mathcal{F}, I_{k-1}^{\leqslant \ell}(\mathcal{F})] \subset I_k^{\leqslant \ell}(\mathcal{F}).$

c. $I_k^{\leqslant \ell+1}(\mathcal{F}) = \mathcal{F}[\mathcal{F}, I_{k-1}^{\leqslant \ell}(\mathcal{F})] + [\mathcal{F}, I_{k-1}^{\leqslant \ell+1}(\mathcal{F})].$

Proof: To prove (a) and (b), we need the following obvious recursion for k, $\ell \geq$ (F), $\ell > 1$:

$$I_k^\ell(\mathcal{F}) = \sum_{i \geqslant 0} \mathcal{F}^{(i)} I_{k-i}^{\ell-1}(\mathcal{F}) \tag{1.3}$$

(With the natural convention that $I_{k'}^{\ell'}(F) = 0$ if $k' < 0$). Then we prove (a) by induction in ℓ If $\ell = 1$, the assertion becomes $F^{(k)} \subset F^{(k-1)}$. Iterating this inclusion and using the inductive hypothesis, we obtain for $\ell > 1$ I

$$I_k^{\ell}(\mathcal{F}) = \sum_{i \geqslant 0} \mathcal{F}^{(i)} I_{k-i}^{\ell-1}(\mathcal{F}) = \mathcal{F} I_k^{\ell-1}(\mathcal{F}) + \sum_{i > 0} \mathcal{F}^{(i)} I_{k-i}^{\ell-1}(\mathcal{F})$$

$$\subset \mathcal{F} I_{k-1}^{\ell-1}(\mathcal{F}) + \sum_{i > 0} \mathcal{F}^{(i-1)} I_{k-i}^{\ell-1}(\mathcal{F}) = \sum_{i \geqslant 0} \mathcal{F}^{(i)} I_{k-i-1}^{\ell-1}(\mathcal{F}) = I_{k-1}^{\ell}(\mathcal{F}).$$

This proves (a)

Prove (b) also by induction in ℓ If $\ell = 1$, the assertion becomes $[F, F^{(k-1)}] \subset F^{(k)}$, which is obvious. Using the inductive hypothesis, we obtain

$$[\mathcal{F}, I_{k-1}^{\ell}(\mathcal{F})] = \sum_{i \geqslant 1} [\mathcal{F}, \mathcal{F}^{(i-1)} I_{k-i}^{\ell-1}(\mathcal{F})] \subset \sum_{i \geqslant 1} [\mathcal{F}, \mathcal{F}^{(i-1)}] I_{k-i}^{\ell-1}(\mathcal{F}) + \mathcal{F}^{(i-1)} [\mathcal{F}, I_{k-i}^{\ell-1}(\mathcal{F})]$$

$$\subset \sum_{i \geqslant 0} \mathcal{F}^{(i)} I_{k-i}^{\ell-1}(\mathcal{F}) = I_k^{\ell}(\mathcal{F}).$$

This proves (b).

Prove (c). Obviously, $I_k^{\leq \ell+1}(F) \supset F[F, I_{k-1}^{\leq \ell+1}(F)] + [F, I_{k-1}^{\leq \ell+1}(F)]$ by (b). Therefore, it suf- fices to prove the opposite inclusion

$$I_k^{\leqslant \ell+1}(\mathcal{F}) \subset \mathcal{F}[\mathcal{F}, I_{k-1}^{\leqslant \ell}(\mathcal{F})] + [\mathcal{F}, I_{k-1}^{\leqslant \ell+1}(\mathcal{F})].$$

We will use the following obvious consequence of (1.3):

$$I_k^{\leqslant \ell+1}(\mathcal{F}) = \sum_{i \geqslant 0} \mathcal{F}^{(i)} I_{k-i}^{\leqslant \ell}(\mathcal{F}).$$

Therefore, it suffices to prove that

$$\mathcal{F}^{(i)} I_{k-i}^{\leqslant \ell}(\mathcal{F}) \subset \mathcal{F}[\mathcal{F}, I_{k-1}^{\leqslant \ell}(\mathcal{F})] + [\mathcal{F}, I_{k-1}^{\leqslant \ell+1}(\mathcal{F})] \tag{1.4}$$

For all i ≥ 0, $\ell \geq 1$, k ≥ 1. We prove (1.4) by induction in all pairs (ℓ ,i) ordered lexicographically. Indeed, suppose that the assertion is proved for all $(\ell', l') < (\ell, i)$. The base of induction is when $\ell = 1$, $i = 0$. Indeed, $I_k^{\leq 1}(F) = F^{(k)}$ for all k and (1.4) becomes $FF^{(k)} \subset F[F, F^{(k-1)}] + [F, I_{k-1}^{\leq 2}(F)]$, which is obviously true since $[F, F^{(k-1)}] = F^{(k)}$. If $\ell \geq 1$, $i > 0$, we obtain, using the Leibniz rule, the following inclusion.

$$\mathcal{F}^{(i)} I_{k-i}^{\leq \ell}(\mathcal{F}) = [\mathcal{F}, \mathcal{F}^{(i-1)}] I_{k-i}^{\leq \ell}(\mathcal{F}) \subset [\mathcal{F}, \mathcal{F}^{(i-1)} I_{k-i}^{\leq \ell}(\mathcal{F})] + \mathcal{F}^{(i-1)}[\mathcal{F}, I_{k-i}^{\leq \ell}(\mathcal{F})].$$

Therefore, by (b)

$$\mathcal{F}^{(i)} I_{k-i}^{\leq \ell}(\mathcal{F}) \subset [\mathcal{F}, \mathcal{F}^{(i-1)} I_{k-i}^{\leq \ell}(\mathcal{F})] + \mathcal{F}^{(i-1)} I_{k+1-i}^{\leq \ell}(\mathcal{F}).$$

Finally, using the inductive hypothesis for $(\ell, i-1)$ and taking into account that $F^{(i-1)} I_{k-i}^{\leq \ell}(F) \subset I_{k-i}^{\leq \ell}(F)$, and, therefore.

$$[\mathcal{F}, \mathcal{F}^{(i-1)} I_{k-i}^{\leq \ell}(\mathcal{F})] \subset [\mathcal{F}, I_{k-1}^{\leq \ell+1}(\mathcal{F})],$$

We obtain the inclusion (1.4).

If $\ell \geq 2$, $i = 0$, then using the inductive hypothesis for all pairs $(\ell - 1, l')$, $l' \geq 0$, we obtain.

$$I_k^{\leq \ell}(\mathcal{F}) = \mathcal{F}[\mathcal{F}, I_{k-1}^{\leq \ell-1}(\mathcal{F})] + [\mathcal{F}, I_{k-1}^{\leq \ell}(\mathcal{F})].$$

Multiplying by F on the left we obtain:

$$\mathcal{F} I_k^{\leq \ell}(\mathcal{F}) = \mathcal{F}^2[\mathcal{F}, I_{k-1}^{\leq \ell-1}(\mathcal{F})] + \mathcal{F}[\mathcal{F}, I_{k-1}^{\leq \ell}(\mathcal{F})] = \mathcal{F}[\mathcal{F}, I_{k-1}^{\leq \ell}(\mathcal{F})]$$

Because $F^2 \subset F$ and $I_{k-i}^{\leq \ell-1}(F) \subset I_{k-i}^{\leq \ell}(F)$. This immediately implies (1.4).

Part (c) is proved. The lemma is proved.

Lemma 1.3. For any $k', k \geq 0$, and any $\ell, \ell \geq 1$ one has.

a. $I_k^{\ell}(\mathcal{F}) I_{k'}^{\ell'}(\mathcal{F}) \subset I_{k+k'}^{\ell+\ell'}(\mathcal{F})$, $I_k^{\leq \ell}(\mathcal{F}) I_{k'}^{\leq \ell'}(\mathcal{F}) \subset I_{k+k'}^{\leq \ell+\ell'}(\mathcal{F})$.
b. $[I_k^{\ell}(\mathcal{F}), I_{k'}^{\ell'}(\mathcal{F})] \subset [\mathcal{F}, I_{k+k'}^{\ell+\ell'-1}(\mathcal{F})]$, $[I_k^{\leq \ell}(\mathcal{F}), I_{k'}^{\leq \ell'}(\mathcal{F})] \subset [\mathcal{F}, I_{k+k'}^{\leq \ell+\ell'-1}(\mathcal{F})]$.

Proof: Part (a) follows from the obvious fact that

$$\left(\mathcal{F}^{(\lambda_1)}\mathcal{F}^{(\lambda_2)}\dots\mathcal{F}^{(\lambda_{\ell_1})}\right)\left(\mathcal{F}^{(\mu_1)}\mathcal{F}^{(\mu_2)}\dots\mathcal{F}^{(\mu_{\ell_2})}\right)\subset I_k^{\ell_1+\ell_2}(\mathcal{F}),$$

Where

$$k=\lambda_1+\lambda_2+\dots+\lambda_{\ell_1}+\mu_1+\mu_2+\dots+\mu_{\ell_2}.$$

Prove (b). First, we prove the first inclusion for $\ell = 1$. We proceed by induction on k. The base of induction, k = 0, is obvious because I_0^1 (F) = F. Assume that the assertion is proved for all $k_1 < k$, i.e., we have:

$$\left[\mathcal{F}^{(k_1)}, I_{k'}^{\ell'}(\mathcal{F})\right]\subset\left[\mathcal{F}, I_{k_1+k'}^{\ell'}(\mathcal{F})\right].$$

Then, using the fact that $F^{(k)} = [F,F^{(k-1)}]$ and the Jacobi identity, we obtain:

$$\left[\mathcal{F}^{(k)}, I_{k'}^{\ell'}(\mathcal{F})\right]=\left[\left[\mathcal{F},\mathcal{F}^{(k-1)}\right], I_{k'}^{\ell'}(\mathcal{F})\right]$$

$$\subset\left[\mathcal{F},\left[\mathcal{F}^{(k-1)}, I_{k'}^{\ell'}(\mathcal{F})\right]\right]+\left[\mathcal{F}^{(k-1)},\left[\mathcal{F}, I_{k'}^{\ell'}(\mathcal{F})\right]\right]$$

$$\subset\left[\mathcal{F},\left[\mathcal{F}, I_{k'+k-1}^{\ell'}(\mathcal{F})\right]\right]+\left[\mathcal{F}^{(k-1)}, I_{k'+1}^{\ell'}(\mathcal{F})\right]\subset\left[\mathcal{F}, I_{k'+k}^{\ell'}(\mathcal{F})\right]$$

By the inductive hypothesis and Lemma 1.2(b). This proves the first inclusion of (b) for $\ell = 1$. Furthermore, we will proceed by induction on ℓ Now $\ell > 1$, assume that the assertion is p

Proved for all $\ell_1 < \ell$, i.e., we have the inductive hypothesis in the form:

$$\left[I_k^{\ell_1}(\mathcal{F}), I_{k'}^{\ell'}(\mathcal{F})\right]\subset\left[\mathcal{F}, I_{k+k'}^{\ell_1+\ell'-1}(\mathcal{F})\right]$$

For all $k,k' \geq 0$.

We need the following useful Jacobi–Leibniz type identity in F:

$$[ab,c]+[bc,a]+[ca,b]=0 \tag{1.5}$$

For all a,b,c ∈ F. (The identity was communicated to the authors by C. Reutenauer and was used in a different context in the recent paper [5].)

Using (1.3) and (1.5) with all a $\in F^{(i)}$, b ∈ $I^{\ell-1}_{k-i}$(F), c ∈ $I^{\ell'}_{k'}$(F), we obtain for all i≥0:

$$[\mathcal{F}^{(i)} I^{\ell-1}_{k-i}(\mathcal{F}), I^{\ell'}_{k'}(\mathcal{F})] \subset [\mathcal{F}^{(i)}, I^{\ell-1}_{k-i}(\mathcal{F}) I^{\ell'}_{k'}(\mathcal{F})] + [I^{\ell-1}_{k-i}(\mathcal{F}), I^{\ell'}_{k'}(\mathcal{F})\mathcal{F}^{(i)}]$$

$$\subset [\mathcal{F}^{(i)}, I^{\ell+\ell'-1}_{k+k'-i}(\mathcal{F})] + [I^{\ell-1}_{k-i}(\mathcal{F}), I^{\ell'+1}_{k'+i}(\mathcal{F})] \subset [\mathcal{F}, I^{\ell+\ell'-1}_{k+k'}(\mathcal{F})]$$

By the already proved (a) and inductive hypothesis. This finishes the proof of the first inclusion of (b). The second inclusion of (b) also follows.

Generalizing (1.3), for any subset S of F denote by $I^{\ell'}_{k'}$ (F,S) the image of Span $S^{\otimes(k+\ell)}$ under the canonical map $F^{\otimes(k+\ell)}$ $I^{\ell'}_{k'}$ (F), i.e.,

$$I^{\ell}_{k}(\mathcal{F}, S) = \sum_{\lambda} S^{(\lambda_1)} S^{(\lambda_2)} \cdots S^{(\lambda_\ell)}, \tag{1.6}$$

$$\lambda = (\lambda_1, \lambda_2, \ldots, \lambda_\ell) \in (\mathbb{Z}_{\geq 0})^\ell$$

Such that

$$\textstyle\sum_{i=1}^{\ell} \lambda_i = k.$$

In particular, I^1_k (F,S) = $S^{(k)}$ and $I^\ell_0 = S^\ell$ The following result is obvious.

Lemma 1.4. Let F be an object of N and S ⊂ F. Then:

(a) For any k ≥0, ℓ ≥2 one has

$$I^{\ell}_{k}(\mathcal{F}, S) = \sum_{i=0}^{k} S^{(i)} I^{\ell-1}_{k-i}(\mathcal{F}, S).$$

(b) For any k′ ,k ≥0, and any ℓ , ℓ ≥ 1 one has:

$$I_k^{\ell}(\mathcal{F},S)I_{k'}^{\ell'}(\mathcal{F},S)\subset I_{k+k'}^{\ell+\ell'}(\mathcal{F},S),\qquad \left[I_k^{\ell}(\mathcal{F},S),I_{k'}^{\ell'}(\mathcal{F},S)\right]\subset I_{k+k'+1}^{\ell+\ell'-1}(\mathcal{F},S).$$

In particular,

$$S^{(i)}I_k^{\ell}(\mathcal{F},S)\subset I_{k+i}^{\ell+1}(\mathcal{F},S),\qquad \left[S^{(i)},I_k^{\ell}(\mathcal{F},S)\right]\subset I_{k+i+1}^{\ell}(\mathcal{F},S). \tag{1.7}$$

N -LIE ALGEBRAS AND N -CURRENT LIE ALGEBRAS

Given objects F and A of Alg, we refer to a morphism ı:F → A in Alg as an F-algebra structure on A (we will also refer to A as an F-algebra). Note that each F-algebra structure on A turns A into an algebra in the category of Fbimodules (i.e., A admits two F-actions F ⊗ A → A, A ⊗ F → A via f ⊗ a → ı(f) · a and a ⊗ f → a ı(f) respectively).

We fix an arbitrary sub-category N of Alg throughout the section. In most cases we take N = Alg.

Definition 2.1: An N -Lie algebra is a triple (F,L,A), where F is an object of N , A is an Falgebra, and L is an F-Lie subalgebra of A, i.e., if L is a Lie subalgebra (under the commutator bracket) of A invariant under the adjoint action of F on A given by (f,a) → ı(f) a-a ı(f) for all f ∈ F, a ∈ A.

A morphism $(F_1,L_1,A_1) \to (F_2,L_2,A_2)$ of N -Lie algebras is a pair (ϕ,ψ), where $\phi : F_1 \to F_2$ is a morphism in N and $\psi : A_1 \to A_2$ is a morphism in Alg such that $\psi(L_1) \subset L_2$ and $\psi \circ \imath_1 = \imath_2 \circ \phi$.

Denote by LieAlg N the category of N -Lie algebras.

For an N -Lie algebra (F,L,A), let Li(F,L,A) := $(F,L^{(i)},A)$, $0 \le i \le j \le 3$, where $L^{(i)}$, i = 1, 2, 3, are given by:

- $L^{(1)}$ is the normalizer Lie algebra of L in A.
- $L^{(2)}$ is the Lie subalgebra of A generated by ı(F) ⊂ A and by the semigroup S = {s ∈ A: s L = L s}.
- L(3) is the Lie subalgebra of A generated by G(ı(F)) ⊂ A, where G is the stabilizer of L in the group Aut $\mathbb{K}$ (A), i.e.,

$$\mathcal{G}=\left\{g\in Aut_{\mathbb{K}}(\mathcal{A}):\ g(\mathcal{L})=\mathcal{L}\right\}. \tag{2.1}$$

The following result is obvious.

Lemma 2.2: For each N -Lie algebra (F,L,A) and i = 1, 2, 3 the triple Li(F,L,A) is also an N -Lie algebra.

Therefore, we can construct a number of new N -Lie algebras by combining the operations Li for a given N -Lie algebra.

Remark 2.3: In general, none of L_i defines a functor $LieAlg_N \to LieAlg_N$. However, for each i = 1, 2, 3 one can find an appropriate subcategory C of $LieAlg_N$ such the restriction of Li to C is a functor $C \to LieAlg_N$

Remark 2.4: The operation L_3 is interesting only when F is noncommutative because for any object (F,L,A) of $LieAlg_N$ such that F is commutative and all automorphisms of A are inner, one obtains $L^{(3)} = \imath(F)$ and therefore, $L_3(F,L,A) = (F,\imath(F),A)$.

Denote by π the natural (forgetful) projection functor $LieAlg_N \to N$ such that $\pi(F,L,A) = F$ and $\pi(\phi,\psi) = \phi$.

Definition 2.5: A noncommutative current Lie algebra (N -current Lie algebra) is a functor $s : N \to LieAlg_N$ such that $\pi \circ s = Id_N$ (i.e., s is a section of π).

Note that if $N = (F,Id_F)$ has only one object F and only the identity arrow Id_F, then the N -current Lie algebra is simply any object of $LieAlg_{Alg}$ of the form (F,L,A). In this case, we will sometimes refer to the Lie algebra L as an F-current Lie algebra.

In principle, we can construct a number of F-current Lie algebras by twisting a given one with operations L_i from Lemma 2.2. However, the study of such "derived" F-current Lie algebras is beyond the scope of the present paper.

In what follows we will suppress the tensor sign in expressions like F ⊗ A and write F A instead. Note that for any object A of Alg1 and any object F of Alg the product F ⊗A is naturally an F-algebra via the embedding $F \to F \cdot A$ $(f \to f\ 1)$.

The following is a first obvious example of N -current Lie algebras.

Lemma 2.6: For any object algebra A of Alg1 and any object F of N define the object $s_A(F) = (F, F \cdot A, F \cdot A)$ of $LieAlg_N$. Then the association F $\rightarrow s_A(F)$ defines a noncommutative current Lie algebra $s_A : N \rightarrow LieAlg_N$.

The main object of our study will be a refinement of the above example. Given an object A of Alg_1, and a subspace $g \subset A$ such that $[g, g] \subset g$ (i.e., g is a Lie subalgebra of A), we say that (g,A) is a compatible pair. For any compatible pair (g,A) and an object F of N , denote by (g,A)(F) the Lie subalgebra of the $F \cdot A = F \otimes A$ (under the commutator bracket) generated by $F \cdot g$, that is, $(g,A)(F) = (F\, g)^{(\bullet)}$ in notation (1.1).

Proposition 2.7: For any compatible pair (g,A) the association

$$\mathcal{F} \mapsto (\mathcal{F}, (\mathfrak{g}, A)(\mathcal{F}), \mathcal{F} \cdot A)$$

Defines the N -current Lie algebra

$$(\mathfrak{g}, A) : \mathcal{N} \rightarrow \mathbf{LieAlg}_{\mathcal{N}}.$$

Proof: It suffices to show that any arrow ϕ in N , i.e., any algebra homomorphism $\phi : F_1 \rightarrow F_2$ defines a homomorphism of Lie algebras $(g,A)(F_1) \rightarrow (g,A)(F_2)$. We need the following obvious fact.

Lemma 2.8: Let A_1, A_2 be objects of Alg and let $\phi : A_1 \rightarrow A_2$ be a morphism in Alg. Let $S_1 \subset A_1$ and $S_2 \subset A_2$ be two subsets such that $\phi(S_1) \subset S_2$. Then the restriction of ϕ to the Lie algebra $S_1^{(\bullet)}$ (in notation (1.1)) is a homomorphism of Lie algebras $S_1^{(\bullet)} \rightarrow S_2^{(\bullet)}$.

Applying Lemma 2.8 with $A_i = F_i A$, $S_i = F_i \cdot g$, i = 1, 2, $\phi = f \otimes id_A : F_1 A \rightarrow F_2 \cdot A$, we obtain a Lie algebra homomorphism $(g,A)(F_1) = (F_1 \cdot g)^{(\bullet)} \rightarrow (F_2 \cdot g)^{(\bullet)} = (g,A)(F_2)$. It remains to show that the action of F on $L = (g,A)(F) = (F \cdot g)^{(\bullet)}$ is stable under the commutator bracket with F. Indeed, $S = F \cdot g$ is invariant under the adjoint action of F on $F \cdot A$. By induction and the Jacobi identity $L = (F \cdot g)^{(\bullet)}$ is also invariant under this action of F. The proposition is proved.

If F is commutative, then (g,A)(F) = F · g is the F-current algebra. Therefore, if F is an arbitrary object of N , the Lie algebra (g,A)(F) deserves a name of the N -current Lie algebra associated with the compatible pair (g,A).

If A = U(g), the universal enveloping algebra of g, then we abbreviate g(F) := (g,U(g))(F). Another natural choice of A is algebra End (V), where V is a faithful g-module. In this case, we will sometimes abbreviate (g,V)(F) := (g,End(V))(F).

The following result provides an estimation of (g,A)(F) from below. Set

$$\langle \mathfrak{g} \rangle = \sum_{k \geqslant 1} \mathfrak{g}^k, \qquad (2.2)$$

I.e. $\langle g \rangle$ is the associative subalgebra of A generated by g.

Proposition 2.9: Let (g,A) be a compatible pair and F be an object of N . Then

(a) $F^{(k)} \cdot g^{k+1} \subset (g,A)(F)$ and $FF^{(k)} \cdot [g, g^{k+1}] \subset (g,A)(F)$ for all $k \geq 0$.
(b) (b) If g is abelian, i.e., $g' = [g, g] = 0$, then

$$(\mathfrak{g}, A)(\mathcal{F}) = \sum_{k \geqslant 0} \mathcal{F}^{(k)} \cdot \mathfrak{g}^{k+1}. \qquad (2.3)$$

(c) If [F,F] = F (i.e., F is perfect as a Lie algebra), then $(g,A)(F) = F \cdot \langle g \rangle$.
Proof: Prove (a). We need the following technical result.

Lemma 2.10: Let (g,A) be a compatible pair. For all $m \geq 2$ denote by

denote by $\tilde{g}^m$ the $\mathbb{K}$ linear span of all power g^m, $g \in g$ them for any $m \geq 2$one has.

$$\widetilde{\mathfrak{g}^m} + (\mathfrak{g}^{m-1} \cap \mathfrak{g}^m) = \mathfrak{g}^m. \qquad (2.4)$$

Proof: Since $g^{i-1}g\, g^{m-i-1} \subset g^{m-1}$ for all $i \leq m-1$, we obtain the following congruence for any $c = (c_1,\ldots,c_m)$, $c_i \in (\mathbb{K} - \{0\})$ and $x = (x_1,\ldots,x_m)$, $x_i \in g$, $i = 1, 2,\ldots,m$:

$$(c_1x_1 + \cdots + c_mx_m)^m \equiv \sum_\lambda \binom{m}{\lambda} c^\lambda x^\lambda \mod (\mathfrak{g}^{m-1} \cap \mathfrak{g}^m).$$

Where the summation is over all partitions $\lambda = (\lambda_1,\ldots,\lambda_m)$ of m and we abbreviated $c^\lambda = C_1^{\lambda_1} \cdots c \lambda m$ m and $x^\lambda = x_1^{\lambda_1} \cdots x_m^{\lambda_m}$ Varying $c = (c_1,\ldots,c_m)$, the above congruence implies that each monomial x^λ belongs to $\tilde{g}^m + (g^{m-1} \cap g^m)$. In particular, taking $\lambda = (1, 1,\ldots, 1)$, we obtain $g^m \subseteq \tilde{g}^m + (g^{m-1} \cap g^m)$. Taking into account that $\tilde{g}^m \subseteq \tilde{g}^m$ we obtain (2.4). The lemma is proved.

We also need the following useful identity in F A:

$$[sE, tF] = st \cdot [E, F] + [s, t] \cdot FE = ts \cdot [E, F] + [s, t] \cdot EF \tag{2.5}$$

For any $s,t \in F$, $E,F \in A$.

We prove the first inclusion of (a) by induction on k. If $k = 0$, one obviously has $F^{(0)} g^1 = F \cdot g \subset (g,A)(F)$. Assume now that $k > 0$. Then for $g \in g$ we obtain by using (2.5):

$$[\mathcal{F} \cdot g, \mathcal{F}^{(k-1)} \cdot g^k] = [\mathcal{F}, \mathcal{F}^{(k-1)}] \cdot g^{k+1} = \mathcal{F}^{(k)} \cdot g^{k+1}$$

Which implies that $F^{(k)}\, \tilde{g}^{k+1} \subset (g,A)(F)$ (in the notation of Lemma 2.10). Using Lemma 2.10, we obtain

$$\mathcal{F}^{(k)} \cdot \widetilde{\mathfrak{g}^{k+1}} \equiv \mathcal{F}^{(k)} \cdot \mathfrak{g}^{k+1} \mod \mathcal{F}^{(k)} \cdot (\mathfrak{g}^k \cap \mathfrak{g}^{k+1}).$$

Taking into account that $F^{(k)} \cdot (g^k \cap g^{k+1}) \subset F^{(k-1)} \cdot g^k \subset (g,A)(F)$ by the inductive hypothesis (here we used the inclusion $F^{(k)} \subset F^{(k-1)}$), the above relation implies that F(k) · gk+1 also belongs to (g,A)(F). This proves the first inclusion of (a). To prove the second inclusion, we compute using (2.5)

$$[\mathcal{F}\cdot\mathfrak{g},\mathcal{F}^{(k-1)}\cdot\mathfrak{g}^k]\equiv\mathcal{F}\mathcal{F}^{(k-1)}\cdot[\mathfrak{g},\mathfrak{g}^k] \mod \mathcal{F}^{(k)}\cdot\mathfrak{g}^{k+1}.$$

Therefore, using the already proved inclusion $F^{(k)} \cdot g^{k+1} \subset (g,A)(F)$, we see that $FF^{(k-1)} \cdot [g, g^k]$ also belongs to (g,A)(F). This finishes the proof of (a).

Prove (b). Clearly, (a) implies that (g,A)(F) contains the right-hand side of (2.3). Therefore, it suffices to prove that the latter space is closed under the commutator. Indeed, since g is abelian, one has

$$[\mathcal{F}^{(k_1)}\cdot\mathfrak{g}^{k_1+1},\mathcal{F}^{(k_2)}\cdot\mathfrak{g}^{k_2+1}]=[\mathcal{F}^{(k_1)},\mathcal{F}^{(k_2)}]\cdot\mathfrak{g}^{k_1+k_2+2}$$

$$\subset\mathcal{F}^{(k_1+k_2+1)}\cdot\mathfrak{g}^{k_1+k_2+2}\subset(\mathfrak{g},A)(\mathcal{F})$$

because $[F^{(k1)}, F^{(k2)}] \subset F^{(k1+k2+1)}$. This finishes the proof of (b).

Prove (c). Since F′ = F, the already proved part (a) implies $F \cdot g^k \subset (g,A)$

(F) for all $k \geq 1$, therefore, $F \cdot \langle g \rangle$

$\subseteq$ (g,A)(F). Since $\langle g \rangle$ is an associative subalgebra of A containing g, we

obtain an opposite inclusion $(g,A)(F) \subseteq F \cdot \langle g \rangle$ This finishes the proof of (c).

The proposition is proved.

Remark 2.11: Proposition 2.9(c) shows that the case when [F,F] = F is not of much interest. This happens, for example, when F is a Weyl algebra or the quantum torus. In these cases a natural anti-involution on F can be taken into account. We will discuss it in a separate paper.

Definition 2.12: We say that a compatible pair (g,A) is of finite type if there exist $m > 0$ such that $g + g^2 + \cdot\cdot + g^m = A$, and we call such minimal m the type of (g,A). If such m does not exists, we say that (g,A) is of infinite type.

Note that (g,A) is of type 1 if and only if g = A, which, in its turn, implies that $(g,A)(F) = F \cdot A$ for all objects F of N . Note also that if $\langle g \rangle = A$ and A is finite-dimensional over $\mathbb{K}$, then (g,A) is always of finite type.

Proposition 2.13: Assume that (g,A) is of type 2, i.e., $g \neq A$ and $g + g^2 = A$. Then

$$(\mathfrak{g}, A)(\mathcal{F}) = \mathcal{F} \cdot \mathfrak{g} + \mathcal{F}' \cdot A + \mathcal{F}\mathcal{F}' \cdot [A, A], \tag{2.6}$$

Where $F' = [F,F]$.

Proof: Since Furthermore, Proposition 2.9(a) guarantees that

$$\mathcal{F} \cdot \mathfrak{g} + \mathcal{F}' \cdot \mathfrak{g}^2 + \mathcal{F}\mathcal{F}' \cdot \left[\mathfrak{g}, \mathfrak{g}^2\right] \subset (\mathfrak{g}, A)(\mathcal{F}).$$

Clearly, $F\,g + F'\,g^2 = F\,g + F' \cdot A$ (because $F' \subset F$). Let us now prove that $[g,A] = [A,A]$. Obviously, $[g,A] \subseteq [A,A]$. The opposite inclusion immediately follows from inclusion $[g^2, g^2] \subseteq [g, g^3]$, which, in its turn follows from (1.5): taking any $a \in g$, $b \in g$, $c \in g^2$ in (1.5), we obtain $[ab,c] \in [g, g^3]$.

Using the equation $[g,A]=[A,A]$ we obtain $FF'\ [A,A] \subset (g,A)(F)$. This proves that (g,A)(F) contains the right-hand side of (2.6).

To finish the proof, it suffices to show that the latter subspace is closed under the commutator. Indeed, abbreviating $A' = [A,A]$, we obtain

$$[\mathcal{F} \cdot \mathfrak{g}, \mathcal{F}' \cdot A] \subset \mathcal{F}\mathcal{F}' \cdot [\mathfrak{g}, A] + [\mathcal{F}, \mathcal{F}'] \cdot A\mathfrak{g} \subset \mathcal{F}\mathcal{F}' \cdot A' + \mathcal{F}' \cdot A,$$

$$[\mathcal{F} \cdot \mathfrak{g}, \mathcal{F}\mathcal{F}' \cdot A'] \subset \mathcal{F}^2\mathcal{F}' \cdot [\mathfrak{g}, A'] + [\mathcal{F}, \mathcal{F}\mathcal{F}'] \cdot A'\mathfrak{g} \subset \mathcal{F}\mathcal{F}' \cdot A' + \mathcal{F}' \cdot A.$$

$$[\mathcal{F}'\cdot A,\mathcal{F}'\cdot A]\subset(\mathcal{F}')^2\cdot A'+[\mathcal{F}',\mathcal{F}']\cdot A^2\subset\mathcal{F}\mathcal{F}'\cdot A'+\mathcal{F}'\cdot A,$$

$$[\mathcal{F}'\cdot A,\mathcal{F}\mathcal{F}'\cdot A']\subset\mathcal{F}'\mathcal{F}\mathcal{F}'\cdot[A,A']+[\mathcal{F}',\mathcal{F}\mathcal{F}']\cdot A'A$$

Because

$$\mathcal{F}'\mathcal{F}\mathcal{F}'\cdot[A,A']\subset\mathcal{F}\mathcal{F}'\cdot A'\subset(\mathfrak{g},A)(\mathcal{F})$$

And

$$[\mathcal{F}',\mathcal{F}\mathcal{F}']\cdot A'A\subset\mathcal{F}'\cdot A\subset(\mathfrak{g},A)(\mathcal{F}).$$

Finally

$$[\mathcal{F}'\mathcal{F}\cdot A',\mathcal{F}\mathcal{F}'\cdot A']\subset(\mathcal{F}'\mathcal{F})^2\cdot[A',A']+[\mathcal{F}'\mathcal{F},\mathcal{F}\mathcal{F}']\cdot(A')^2$$

because (F F)^2 [A′ ,A] ⊂ F′ F A′ and [F′ F,FF′]·[A,A]^2 ⊂ F′ A. The proposition is proved.

For any k-vector space V and any object F of N we abbreviate sl(V,F):= (sl(V), End(V))(F) and gl(V,F) := (End(V),End(V))(F) = F · End(V). The following result shows that for n = dimV≥2 the commutator Lie algebra sl_n(F) = [gl_n(F), gl_n(F)] is, in fact, $_s$l(V,F).

Corollary 2.14: Let V be a finite-dimensional k-vector space such that dimV > 1. Then (g,A) = (sl(V),End(V)) is of type 2 and

$$sl(V,\mathcal{F})=\mathcal{F}'\cdot 1+\mathcal{F}\cdot sl(V).$$

Hence $_s$l(V,F) is the set of all X ∈ gl(V,F) such that Tr(X) ∈ F′ = [F,F] (where Tr:gl(V,F) = F · End(V) → F is the trivial extension of the ordinary trace End(V) → $\mathbb{K}$).

Proof: Let us prove that the pair (g,A) = (sl(V),End(V)) is of type 2, i.e., $_s$l(V) + sl(V)^2 = End(V). It suffices to show that 1 ∈ sl(V)^2. To prove it, choose a basis $e_1,\ldots,e_n$ in V so that V ≅ k^n, sl(V) ≅ $sl_n(\mathbb{K})$ and A = End(V) ≅ $M_n(\mathbb{K})$. Indeed, for any indices i ≠ j both E_{ij} and E_{ji} belong to sl(V), and $E_{ij}E_{ji} = E_{ii} \in sl(V)^2$. Therefore, $1 = \sum_{i=1}^{n} E_{ii}$ also belongs to $sl(V)^2$. Applying Proposition 2.13 and using the obvious fact that $[A,A] = sl(V)$, we obtain

$$sl(V,\mathcal{F}) = \mathcal{F} \cdot sl(V) + \mathcal{F}' \cdot A + \mathcal{F}\mathcal{F}'[A,A] = \mathcal{F} \cdot sl(V) + \mathcal{F}' \cdot 1.$$

This proves the first assertion. The second one follows from the obvious fact that the trace Tr:F · End(V) → F is the projection to the second summand of the direct sum decomposition

$$\mathcal{F} \cdot End(V) = \mathcal{F} \cdot sl(V) + \mathcal{F}' \cdot 1.$$

The corollary is proved.

We can construct more pairs of type 2 as follows. Let V be a $\mathbb{K}$-vector space and $\Phi : V \times V \to \mathbb{K}$ be a bilinear form on V . Denote by o(Φ) the orthogonal Lie algebra of Φ, i.e.,

$$o(\Phi) = \{M \in End(V) : \Phi(M(u), v) + \Phi(u, M(v)) = 0 \ \forall u, v \in V\}.$$

Denote by $K_\Phi \subset V$ the sum of the left and the right kernels of Φ (if Φ is symmetric or skewsymmetric, then K_Φ is the left kernel of Φ). Finally, denote by End (V,K_Φ) the parabolic subalgebra of End(V) which consists of all M ∈ End(V) such that $M(K_\Phi) \subset K_\Phi$. Clearly, $o(\Phi) \subset End(V,K_\Phi)$, i.e., $(o(\Phi),End(V,K_\Phi))$ is a compatible pair. For any object F of N we abbreviate $o(\Phi,F) := (o(\Phi),End(V,K_\Phi))(F)$.

Denote by $sl(V,K_\Phi)$ the set of all M in $End(V,K_\Phi)$ such that Tr(M) = 0 and $Tr(M_{K\Phi}) = 0$, where $M_{K\Phi} : K_\Phi \to K_\Phi$ is the restriction of M to K_Φ and $1K_\Phi \in End(V,K_\Phi)$ is any element such that $\mathbb{K} \cdot 1 + \mathbb{K} \cdot 1K_\Phi + sl(V,K_\Phi) = End(V,K_\Phi)$. If $K_\Phi = 0$, we set $1K_\Phi = 0$.

Corollary 2.15: Let V = 0 be a finite-dimensional k-vector space and Φ be a symmetric or skew-symmetric bilinear form on V . Then (o(Φ),End(V,KΦ)) is of type 2 and

$$o(\Phi,\mathcal{F}) = \mathcal{F}\cdot o(\Phi) + \mathcal{F}'\cdot 1 + \mathcal{F}'\cdot 1_K + (\mathcal{F}\mathcal{F}' + \mathcal{F}')\cdot sl(V, K_\Phi). \tag{2.7}$$

Proof: We will write K instead of KΦ. First prove that (g,A) = (o(Φ),End(V,K)) is of type 2. We pass to the algebraic closure of the involved objects, i.e., replace both V and K with $\overline{V} = \overline{\mathbb{K}}\cdot\overline{V} = \overline{\mathbb{K}}\otimes V$, $\bar{k} = \overline{\mathbb{K}}\cdot K$, etc., where $\overline{\mathbb{K}}$ is the algebraic closure of $\mathbb{K}$. Using the obvious fact that $\overline{U+U'} = \overline{U} + \overline{U}$ and $\overline{U\otimes U'} = \overline{U}\otimes\overline{U}$ for any subspaces of End(V) and $\overline{0(\Phi)} = 0(\overline{\Phi})$, we see that it suffices to show that the pair $(0(\overline{\Phi}), \mathrm{End}(\overline{V}, \bar{k}))$ is of type 2.

Furthermore, without loss of generality we consider the case when K = 0, i.e., form Φ is nondegenerate. One can do it by using the block matrix decomposition with respect to a choice of compliment of K over which Φ is nondegenerate.

We will prove the lemma when dimV > 2 and leave the rest to the reader. If Φ is symmetric, one can choose a basis of $\overline{V}$ so that $\overline{V} \cong \overline{\mathbb{K}^n}$, and $\overline{\varnothing}$ is the standard dot product on $\overline{\mathbb{K}^n}$. In this case $o(\overline{\varnothing})$ is $on(\overline{\mathbb{K}})$, the Lie algebra of orthogonal matrices, which is generated by all elements $E_{ij} - E_{ji}$ where E_{ij} is the corresponding elementary matrix. Using the identity $(E_{ij} - E_{ji})^2 = -(E_{ii} + E_{jj})$ for $i \neq j$, we see that $_{0n}(\overline{\mathbb{K}})^2$ contains all diagonal matrices. Furthermore, if i,j,k are pairwise distinct indices then $(E_{ij} - E_{ji})(E_{jk} - E_{kj}) = E_{ik}$. Thus we have shown that $on(\overline{\mathbb{K}})^2 = M_n(\overline{\mathbb{K}}) = \mathrm{End}(\overline{V}, \bar{k})$. Therefore, $on(\mathbb{K})^2 = M_n(\mathbb{K}) = \mathrm{End}(V,K)$. This proves the assertion for the symmetric Φ.

If Φ is skew-symmetric and nondegenerate, then n = 2m and one can choose a basis of $\overline{V}$ such that V is identified with $\overline{\mathbb{K}^n}$ and $o(\overline{\varnothing})$ is identified with the symplectic Lie algebra $sp_{2m}(\overline{\mathbb{K}})$. Recall that a basis in $sp_{2m}(\overline{\mathbb{K}})$ can be chosen as follows. It consists of elements $E_{ij} - E_{j+m,i+m,}$ $E_{i,m+j} + E_{j,m+i}$, $E_{m+i,j} + E_{m+j,i}$, for $i,j \leq m$. Using the identity $(E_{i,i+m} + E_{i+m,i})^2 = E_{ii} + E_{i+m,i+m}$ and the fact that $(E_{ii} - E_{i+m,i+m}) \in sp_{2m}(\mathbb{K})$, we see that all diagonal matrices belong $sp_{2m}(\overline{\mathbb{K}}) + sp_{2m}(\overline{\mathbb{K}})^2$.

Also, the identity $(E_{ii} - E_{i+m,i+m})(E_{ij} - E_{j+m,i+m}) = E_{ij}$ for $i \neq j$ implies that $E_{ij} \in sp_{2m}(\overline{\mathbb{K}})^2$ for all i≤,jm. Similarly, one can prove that $E_{ij} \in sp_{2m}(\overline{\mathbb{K}})^2$ for $i,j \leq m$.

Furthermore, the identity $(E_{ii} - E_{i+m,i+m})(E_{i\ell} + E_{i+m,\ell-m}) = E_{i\ell} - E_{i+m,\ell-m}$ implies that $sp_{2m}(\overline{\mathbb{K}}) + sp_{2m}(\overline{\mathbb{K}})^2$ contains all E_{ik} for $i \leq m, k > m$ and for $i > m, k \leq m$. Thus we have shown that $sp_{2m}(\overline{\mathbb{K}}) + sp_{2m}(\overline{\mathbb{K}})^2 = M_n(\overline{\mathbb{K}}) = End(\overline{V}, \overline{k})$. Therefore, $sp_{2m}(\mathbb{K}) + sp_{2m}(\mathbb{K})^2 = M_n(\mathbb{K}) = End(V,K)$. This proves the assertion for the skew-symmetric Φ. Prove (2.7) now. We abbreviate A = End(V,K). Recall that $[A,A] = {}_{sl}(V,K)$ and, if $K \neq \{0\}$, then $\mathbb{K}\,1 + {}_{sl}(V,K)$ is of codimension 1 in A, i.e., 1K always exists. Therefore, applying Proposition 2.13, we obtain

$$\begin{aligned} o(\Phi, \mathcal{F}) &= \mathcal{F} \cdot o(\Phi) + \mathcal{F}' \cdot A + \mathcal{F}\mathcal{F}'[A, A] \\ &= \mathcal{F} \cdot o(\Phi) + \mathcal{F}' \cdot 1 + \mathcal{F}' \cdot 1_K + (\mathcal{F}\mathcal{F}' + \mathcal{F}') \cdot sl(V, K). \end{aligned}$$

This finishes the proof of Corollary 2.15.

Note that our Lie algebras ${}_{o}(F)$ and sp(F) do not coincide with the usual orthogonal and symplectic Lie algebras, which are defined when the ring F possesses an involution.

UPPER BOUNDS OF N -CURRENT LIE ALGEBRAS

For any compatible pair (g,A) define two subspaces $\widetilde{(g,A)}$ (F) and $\overline{(g,A)}$ (F) of F A by:

$$\widetilde{(\mathfrak{g}, A)}(\mathcal{F}) = \mathcal{F} \cdot \mathfrak{g} + \sum_{k \geqslant 1} I_k(\mathcal{F}) \cdot [\mathfrak{g}, \mathfrak{g}^{k+1}] + [\mathcal{F}, I_{k-1}(\mathcal{F})] \cdot \mathfrak{g}^{k+1}. \tag{3.1}$$

Where $I_k(F)$ is defined in (1.3); and

$$\overline{(\mathfrak{g}, A)}(\mathcal{F}) = \mathcal{F} \cdot \mathfrak{g} + \sum I_{k_1}^{\ell_1+1} I_{k_2}^{\ell_2+1} \cdot [J_{\ell_1}^{k_1+1}, J_{\ell_2}^{k_2+1}] + [I_{k_1}^{\ell_1+1}, I_{k_2}^{\ell_2+1}] \cdot J_{\ell_2}^{k_2+1} J_{\ell_1}^{k_1+1}, \tag{3.2}$$

where the summation is over all $k_1, k_2 \geq 0$, $\ell_1, \ell_2 \geq 0$, and we abbreviated $I_k^\ell := I_k^\ell(F)$, $J_k^\ell := I_k^\ell(A, g)$ in notation (1.6).

We will refer to $\widetilde{(g,A)}(F)$ as the upper bound of $\overline{(g,A)}(F)$ and to $\widetilde{(g,A)}(F)$ as the refined upper bound of (g,A)(F).

It is easy to see that the assignments $F \mapsto \widetilde{(g,A)}(F)$ and $F \mapsto \overline{(g,A)}(F)$ are functors $\widetilde{(g,A)}$ and $\overline{(g,A)}$ from N to the category $\mathrm{Vect}_{\mathbb{K}}$ of $\mathbb{K}$-vector spaces.

The following lemma is obvious.

Lemma 3.1: If (g,A) is a compatible pair of type m (see Definition 2.12), then

$$\widetilde{(\mathfrak{g}, A)}(\mathcal{F}) = \mathcal{F} \cdot \mathfrak{g} + \sum_{k=1}^{m-1} I_k(\mathcal{F}) \cdot [\mathfrak{g}, \mathfrak{g}^{k+1}] + [\mathcal{F}, I_{k-1}(\mathcal{F})] \cdot \mathfrak{g}^{k+1}. \tag{3.3}$$

Now we formulate the main result of this section, which explains our terminology and proves that both $\widetilde{(g,A)}$ and $\overline{(g,A)}(F)$ define N-current Lie algebras $N \to \mathrm{LieAlg}_N$.

Theorem 3.2: For any compatible pair (g,A) and any object F of N one has:

a. The subspace $\widetilde{(g,A)}(F)$ is a Lie subalgebra of F A.
b. The subspace $\overline{(g,A)}(F)$ is a Lie subalgebra of F A.

c. $(g,A)(F) \subseteq \overline{(g,A)}(F) \subseteq \widetilde{(g,A)}(F)$.

Proof: We need the following two lemmas.

This is an obvious consequence of (1.5).

For any subsets X and Y of an object A of Alg and $\varepsilon \in \{0, 1\}$ denote

$$X \bullet_\varepsilon Y := \begin{cases} X \cdot Y & \text{if } \varepsilon = 0, \\ [X, Y] & \text{if } \varepsilon = 1. \end{cases}$$

Lemma 3.4: Let Γ be an abelian group and let A and F be objects of Alg. Assume that $E_\alpha \subset F$ and $B_\alpha \subset A$ are two families of subspaces labeled by Γ such that

$$E_\alpha \bullet_\varepsilon E_\beta \subseteq E_{\alpha+\beta+\varepsilon\cdot v}, \qquad B_\beta \bullet_\varepsilon B_\alpha \subseteq B_{\alpha+\beta-\varepsilon\cdot v} \tag{3.4}$$

for all $\alpha,\beta \in \Gamma$, $\varepsilon \in \{0, 1\}$, where v is a fixed element of Γ. Then for any $\alpha_0 \in \Gamma$ the subspace

$$\mathfrak{h} = E_{\alpha_0} \cdot B_{\alpha_0+v} + \sum_{\alpha,\beta\in\Gamma,\, \varepsilon\in\{0,1\}} (E_\alpha \bullet_{1-\varepsilon} E_\beta) \cdot (B_{\beta+v} \bullet_\varepsilon B_{\alpha+v})$$

Is a Lie subalgebra of $F \cdot A = F \otimes A$.

Proof: Eq. (2.5) implies that

$$[E \cdot B, E' \cdot B'] \subset (E \bullet_{1-\delta} E') \cdot (B' \bullet_\delta B)$$

for each $\delta \in \{0, 1\}$.

(i) Set $E = E_\alpha \bullet_{1-\varepsilon} E_\beta$, $B = B_{\beta+v} \bullet_\varepsilon B_{\alpha+v}$, $E' = E_{\alpha'} \bullet_{1-\varepsilon'} E_{\beta'}$, $B' = B_{\beta'+v} \bullet_{\varepsilon'} B_{\alpha'+v}$.

Define

$\alpha''\alpha + \beta + (1-\varepsilon).v$ and $\beta'' = \alpha' + (1-\varepsilon'), v$.

Taking into account that $E \subseteq E_{\alpha''}$, $E' \subseteq E_{\beta''}$, $B \subseteq B_{\alpha''+v,}$ and $B' \subseteq B_{\beta''+v}$ by (3.4), we obtain for each $\delta \in \{0, 1\}$:

$$[E \cdot B, E' \cdot B'] \subset (E_{\alpha''} \bullet_{1-\delta} E_{\beta''}) \cdot (B_{\beta''+v} \bullet_{\delta} B_{\alpha''+v}) \subset \mathfrak{h}.$$

(ii) Set $E = E_{\alpha 0}$, $B = B_{\alpha 0+v,}$ $E' = E_{\alpha' \bullet 1-\varepsilon'} E_{\beta'}$ $B' = B_{\beta'+v \bullet \varepsilon} B_{\alpha+v.}$ Define β'' as above. Taking into account that $E' \subseteq E_{\beta''}$ and $B' \subseteq B_{\beta''+v}$ by (3.4), where $\beta'' = \alpha' + \beta' + (1 - \varepsilon') \cdot v$, we obtain for each $\delta \in \{0, 1\}$:

$$[E \cdot B, E' \cdot B'] \subset (E_{\alpha_0} \bullet_{1-\delta} E_{\beta''}) \cdot (B_{\beta''+v} \bullet_{\delta} B_{\alpha_0+v}) \subset \mathfrak{h}.$$

(ii) Taking $E = E' = E_{\alpha 0}$, $B = B' = B_{\alpha 0+v'}$ we obtain for each $\delta \in \{0, 1\}$:

$$[E \cdot B, E' \cdot B'] \subset (E_{\alpha_0} \bullet_{1-\delta} E_{\alpha_0}) \cdot (B_{\alpha_0+v} \bullet_{\delta} B_{\alpha_0+v}) \subset \mathfrak{h}.$$

The lemma is proved.

Now we are going to prove the theorem part-by-part.

Prove (a). Using (2.5), we obtain

$$[\mathcal{F} \cdot \mathfrak{g}, \mathcal{F} \cdot \mathfrak{g}] \subset \mathcal{F}^2 \cdot [\mathfrak{g}, \mathfrak{g}] + [\mathcal{F}, \mathcal{F}] \cdot \mathfrak{g}^2 \subset \widetilde{(\mathfrak{g}, A)}(\mathcal{F})$$

Because $F^2 \subset F$, $[g, g] \subset g$, and $I_0(F) = F$. Furthermore,

$$[\mathcal{F} \cdot \mathfrak{g}, I_k(\mathcal{F}) \cdot [\mathfrak{g}, \mathfrak{g}^{k+1}]] \subset \mathcal{F} I_k(\mathcal{F}) \cdot [\mathfrak{g}, [\mathfrak{g}, \mathfrak{g}^{k+1}]] + [\mathcal{F}, I_k(\mathcal{F})] \cdot [\mathfrak{g}, \mathfrak{g}^{k+1}]\mathfrak{g} \subset \widetilde{(\mathfrak{g}, A)}(\mathcal{F})$$

Because $FI_k(F) \subset I_k(F)$, $[g,[g, g^{k+1}]] \subset [g, g^{k+1}]$, and $[g, g^{k+1}]g \subset g^k+2$. Finally, set $J_k := [F,I_k-1(F)]$. Since $I_k-_1(F)$ is a two-sided ideal in F, we have $FJ_k \subset FI_{k-1}(F) \subset I_{k-1}(F)$. Lemma 1.2(b) taken with $\ell = \infty$, implies $J_k \subset I_k(F)$. Therefore, $[F,J_k] \subset [F,I_k(F)]$ and

$$[\mathcal{F} \cdot \mathfrak{g}, J_k \cdot \mathfrak{g}^{k+1}] \subset \mathcal{F} \cdot J_k \cdot [\mathfrak{g}, \mathfrak{g}^{k+1}] + [\mathcal{F}, J_k] \cdot \mathfrak{g}^{k+2} \subset \widetilde{(\mathfrak{g}, A)}(\mathcal{F}).$$

Note that for any $k,m \geq 1$ one has:

$$[I_k(\mathcal{F})\cdot \mathfrak{g}^{k+1}, I_m(\mathcal{F})\cdot \mathfrak{g}^{m+1}] \subset I_k(\mathcal{F})I_m(\mathcal{F})\cdot[\mathfrak{g}^{k+1},\mathfrak{g}^{m+1}] + [I_k(\mathcal{F}), I_m(\mathcal{F})]\cdot \mathfrak{g}^{k+m+2} \subset \widetilde{(\mathfrak{g}, A)}(\mathcal{F})$$

because $I_k(F)I_m(F) \subset I_{k+m}(F)$ by Lemma 1.3(a), $[g^{k+1}, g^{m+1}] \subset [g, g^{k+m}]$ by Lemma 3.3, and $[I_k(F), I_m(F)] \subset [F, I_{k+m}-1(F)]$ by Lemma 1.3(b) taken with $\ell = \infty$. Therefore, taking into account that

For I_r standing for any of the spaces $I_r(F)$ $[g, g^{r+1}], [F, Ir-1(F)] \cdot g^{r+1}$ we finish the proof of (a).

Prove (b). Taking in Lemma 3.4: $\Gamma = \mathbb{Z}^2$, $\alpha = (k, +1) \in Z2$, $v = (1,-1)$, one can see that E_α equals to $I_k^{\ell+1}$ (F) if k, $\ell \geq 0$ and zero otherwise. Also, $B_{\alpha+v}$ equals to $I_\ell^{k+1}(A, g)$ if k, $\ell \geq 0$ and zero otherwise. Lemma

Lemma 1.4 implies that (3.4) holds for all $\alpha, \beta \in \mathbb{Z}^2$, $\varepsilon \in \{0, 1\}$. Therefore, applying Lemma 3.4 with $\alpha_0 = (0, 1)$, we finish the proof of the assertion that $\overline{(g, A)}$ (F) is a Lie subalgebra of F A. Prove (c). The first inclusion $\overline{(g, A)}$ (F) $\subset$ $\overline{(g, A)}$ (F) is obvious because $F \cdot g \subset \overline{(g, A)}$ (F) and (g,A)(F) is a Lie subalgebra of F A.

Let us prove the second inclusion (g,A)(F) $\subset$ $\overline{(g, A)}$ (F) of (c).

Rewrite the result of Lemma 1.3 (with $\ell_1 = \ell_2 = \infty$) in the form of (3.4) as:

$$I_{k_1}^{\ell_1+1}(\mathcal{F}) \bullet_{1-\varepsilon} I_{k_2}^{\ell_2+1}(\mathcal{F}) \subset I_{k_1}(\mathcal{F}) \bullet_{1-\varepsilon} I_{k_2}(\mathcal{F}) \subset \begin{cases} I_{k_1+k_2}(\mathcal{F}) & \text{if } \varepsilon = 1, \\ [\mathcal{F}, I_{k_1+k_2-1}(\mathcal{F})] & \text{if } \varepsilon = 0. \end{cases}$$

Using the obvious inclusion $J_k^{\ell+1} = I_k^{\ell+1}(A, g) \subset g^{k+1}$ for all k, $\ell \geq 0$ and Lemma 3.3, we obtain

$$J_{\ell_2}^{k_2+1} \bullet_\varepsilon J_{\ell_1}^{k_1+1} \subset \mathfrak{g}^{k_2+1} \bullet_\varepsilon \mathfrak{g}^{k_1+1} \subset \begin{cases} \mathfrak{g}^{k_1+k_2+2} & \text{if } \varepsilon = 0, \\ [\mathfrak{g}, \mathfrak{g}^{k_1+k_2+1}] & \text{if } \varepsilon = 1 \end{cases}$$

For all $k_1, k_2, \ell_1, \ell_2 \geq 0$, $\varepsilon \in \{0, 1\}$. Therefore, we get the inclusion:

$$(I_{k_1}^{\ell_1+1} \bullet_{1-\varepsilon} I_{k_2}^{\ell_2+1}) \cdot (J_{\ell_2}^{k_2+1} \bullet_{\varepsilon} J_{\ell_1}^{k_1+1}) \subset \begin{cases} I_{k_1+k_2}(\mathcal{F}) \cdot [\mathfrak{g}, \mathfrak{g}^{k_1+k_2+1}] & \text{if } \varepsilon = 1 \\ [\mathcal{F}, I_{k_1+k_2-1}(\mathcal{F})] \cdot \mathfrak{g}^{k_1+k_2+2} & \text{if } \varepsilon = 0 \end{cases} \subset \widetilde{(\mathfrak{g}, A)}(\mathcal{F}).$$

This proves the inclusion $\overline{(g,A)}$ (F) $\subset$ $\widetilde{(g,A)}$ (F) and finishes the proof of (c).

Therefore, Theorem 3.2 is proved.

Now we will refine Theorem 3.2 by introducing a natural filtration on each Lie algebra involved and proving the "filtered" version of the theorem.

For any compatible pair (g,A), any object F of N , and each $m \geq 1$ we define the subspaces F $\langle g \rangle_m$ $(g,A)_m$(F), $\widetilde{(g,A)}_m$(F) and $\overline{(g,A)}_m$(F) of F A by:

$$\mathcal{F} \cdot \langle \mathfrak{g} \rangle_m = \sum_{1 \leqslant k \leqslant m} \mathcal{F} \cdot \mathfrak{g}^k,$$

$$(\mathfrak{g}, A)_m(\mathcal{F}) = \sum_{0 \leqslant k < m} (\mathcal{F} \cdot \mathfrak{g})^{(k)}, \tag{3.5}$$

$$\widetilde{(\mathfrak{g}, A)}_m(\mathcal{F}) = \mathcal{F} \cdot \mathfrak{g} + \sum_{1 \leqslant k < m} I_k^{\leqslant m-k}(\mathcal{F}) \cdot [\mathfrak{g}, \mathfrak{g}^{k+1}] + [\mathcal{F}, I_{k-1}^{\leqslant m-k}(\mathcal{F})] \cdot \mathfrak{g}^{k+1}, \tag{3.6}$$

Where $I_k^{\leq \ell}$ (F) is defined in (1.3) and

$$\overline{(\mathfrak{g}, A)}_m(\mathcal{F}) = \mathcal{F} \cdot \mathfrak{g} + \sum I_{k_1}^{\ell_1+1} I_{k_2}^{\ell_2+1} \cdot [J_{\ell_1}^{k_1+1}, J_{\ell_2}^{k_2+1}] + [I_{k_1}^{\ell_1+1}, I_{k_2}^{\ell_2+1}] \cdot J_{\ell_2}^{k_2+1} J_{\ell_1}^{k_1+1}, \tag{3.7}$$

where the summation is over all $k_1, k_2 \geq 0$, $\ell_1, \ell_2 \geq 0$ such that $k_1 + k_2 + \ell_1 + \ell_2 + 2 \leq m$, and we abbreviated $I_k^\ell := I_k^\ell$ (F), $J_k^\ell := I_k^\ell$ (A, g) in the notation (1.6).

Recall that a Lie algebra $h = (h_1 \subset h_2 \subset \cdots)$ is called a filtered Lie algebra if $[h_{k1}, h_{k2}] \subset h_{k1+k2}$ for all $k_1, k_2 \geq 0$.

Taking into account that $[g_{k_1+1}, g_{k2+1}] \subset g_{k1+k2+1}$, one can see that $h_m = F\langle g \rangle_m, m \geq 1$, defines an increasing filtration on the Lie algebra $F \cdot \langle g \rangle$ (where $\langle g \rangle$ is as in (2.2)).

The following result is a filtered version of Theorem 3.2.

Theorem 3.5: For any compatible pair (g,A) and an object F of N

a. $\widetilde{(g, A)}$ (F) is a filtered Lie subalgebra of $F^{\langle g \rangle}$

b. $\overline{(g, A)}$ (F) is a filtered Lie subalgebra of $F \cdot^{\langle g \rangle}$

c. There is a chain of inclusions of filtered Lie algebras:

$$(\mathfrak{g}, A)(\mathcal{F}) \subseteq \overline{(\mathfrak{g}, A)}(\mathcal{F}) \subseteq \widetilde{(\mathfrak{g}, A)}(\mathcal{F}).$$

The proof of Theorem 3.5 is almost identical to that of Theorem 3.2.

PERFECT PAIRS AND ACHIEVABLE UPPER BOUNDS

Below we lay out some sufficient conditions on the compatible pair (g,A) which guarantee that the upper bounds are achievable.

Definition 4.1: We say that a compatible pair (g,A) is perfect if

$$[\mathfrak{g}, \mathfrak{g}^k]\mathfrak{g} + (\mathfrak{g}^k \cap \mathfrak{g}^{k+1}) = \mathfrak{g}^{k+1} \tag{4.1}$$

Definition 4.2: We say that a Lie algebra $\bar{g}$ over an algebraically closed field $\overline{\mathbb{K}}$ is strongly graded if there exists an element $h_0 \in \bar{g}$ such that:

(i) The operator ad h_0 on $\bar{g}$ is diagonalizable, i.e.,

$$\bar{\mathfrak{g}} = \bigoplus_{c \in \bar{\mathbb{k}}} \bar{\mathfrak{g}}_c, \tag{4.2}$$

Where $\bar{g}c \subset \bar{g}$ is an eigenspace of ad h_0 with the eigenvalue c

(ii) The nullspace g_0 of ad h_0 is spanned by $[\bar{g}c, \bar{g}_{-c}]$, $c \in \bar{\mathbb{K}} \setminus \{0\}$.

The class of strongly graded Lie algebras is rather large; it includes all semisimple and Kac– Moody Lie algebras, as well as the Virasoro algebra.

Main Theorem 4.3: Let (g,A) be a compatible pair. Then

(a) If (g,A) is perfect, then for any object F of N one has

$$(\mathfrak{g}, A)(\mathcal{F}) = \widetilde{(\mathfrak{g}, A)}(\mathcal{F}),$$

i.e., the N -current Lie algebras (g,A), $\widetilde{(g, A)}$: N → LieAlg_N are equal.

(b) If $g = \bar{\mathbb{K}} \otimes g$ is strongly graded, then (g,A) is perfect.

(c) If g is semisimple over $\mathbb{K}$, then for any object F of N one has

$$(\mathfrak{g}, A)(\mathcal{F}) = \mathcal{F} \cdot \mathfrak{g} + \sum_{k \geqslant 2} I_{k-1}(\mathcal{F}) \cdot (\mathfrak{g}^k)_+ + [\mathcal{F}, I_{k-2}(\mathcal{F})] \cdot Z_k(\mathfrak{g}), \tag{4.3}$$

Where $(g^k)+ = [g, g^k]$ is the "centerless" part of g^k , $Z_k(g) = Z(\langle g \rangle) \cap g^k$, and $Z(\langle g \rangle)$is the center of $\langle g \rangle = \Sigma_{k \geq 1} g^k$.

(a) Assume that for some k ≥ 1 one has

$I \cdot [\mathfrak{g}, \mathfrak{g}^k] \subset (\mathfrak{g}, A)(\mathcal{F})$

where I is a left ideal in F. Then:

$$[\mathcal{F}, I] \cdot [\mathfrak{g}, \mathfrak{g}^k]\mathfrak{g} \subset (\mathfrak{g}, A)(\mathcal{F}). \tag{4.4}$$

(b) Assume that for some k≥ 1 one has

$$J \cdot \mathfrak{g}^k \subset (\mathfrak{g}, A)(\mathcal{F})$$

where J is a subset of F such that [F,J] ⊂ J . Then:

$$[\mathcal{F}, J] \cdot \mathfrak{g}^{k+1} + (\mathcal{F}J + J) \cdot [\mathfrak{g}, \mathfrak{g}^k] \subset (\mathfrak{g}, A)(\mathcal{F}) \tag{4.5}$$

Proof: Prove (a). Indeed,

$$[\mathcal{F} \cdot \mathfrak{g}, I \cdot [\mathfrak{g}, \mathfrak{g}^k]] \equiv [\mathcal{F}, I] \cdot [\mathfrak{g}, \mathfrak{g}^k]\mathfrak{g} \mod \mathcal{F}I \cdot [\mathfrak{g}, [\mathfrak{g}, \mathfrak{g}^k]].$$

Since FI ⊂ I and [g,[g, g^k]] ⊂ [g, g^k], and, therefore, FI ·[g,[g, g^k]] ⊂ I [g, g^k] ⊂ (g,A)(F), the above congruence implies that [F,I]·[g, g^k]g also belongs to (g,A)(F). This proves (a). Prove (b). For any g ∈ g we obtain:

$$[\mathcal{F} \cdot g, J \cdot g^k] = [\mathcal{F}, J] \cdot g^{k+1}$$

Which implies that [F,J] $\widetilde{g^{k+1}}$ ⊂ (g,A)(F) (in the notation of Lemma 2.10). Using Lemma 2.10, we obtain

$$[\mathcal{F}, J] \cdot \widetilde{\mathfrak{g}^{k+1}} \equiv [\mathcal{F}, J] \cdot \mathfrak{g}^{k+1} \mod [\mathcal{F}, J] \cdot (\mathfrak{g}^k \cap \mathfrak{g}^{k+1}).$$

Taking into account that [F,J]·(g^k ∩g^{k+1}) ⊂ [F,J]·g^k ⊂ (g,A)(F), the above formula implies that [F,J] · g^{k+1} also belongs to (g,A)(F). Furthermore,

$$[\mathcal{F} \cdot \mathfrak{g}, J \cdot \mathfrak{g}^k] \equiv \mathcal{F}J \cdot [\mathfrak{g}, \mathfrak{g}^k] \mod [\mathcal{F}, J] \cdot \mathfrak{g}^{k+1}.$$

Therefore, using the already proved inclusion [F,J]· g^{k+1} ⊂ (g,A)(F), we see that FJ ·[g, g^k] also belongs to (g,A)(F). Finally, using the fact that [g,

$g^k] \subset g^k$, we obtain $J \cdot [g, g^k] \subset J \cdot g^k \subset (g,A)(F)$. This proves (b).

Proposition 4.4 is proved.

Lemma 4.5: Let (g,A) be a compatible pair. Assume that $h_0 \in \bar{g} = \overline{\mathbb{K}} \otimes g$ is such that ad h_0 is diagonalizable, i.e., has a decomposition (4.2). Then:

a. For each $k \geq 1$ and each $\underline{c} = (c1,\ldots,c_{k+1}) \in \overline{\mathbb{K}}^{k+1} \setminus \{0\}$ the subspace $\bar{g}_{c1} \cdots \bar{g}_{ck+1}$ of $\bar{g}^{k+1}$ belongs to $[\bar{g}, \bar{g}^k]\bar{g} + (\bar{g}^k \cap \bar{g}^{k+1})$.
b. If $\bar{g} = \overline{\mathbb{K}} \otimes g$ is a strongly graded Lie algebra, then one has (in the notation of Definition 4.2):

$$\bar{\mathfrak{g}}_0^{k+1} \subset [\bar{\mathfrak{g}}, \bar{\mathfrak{g}}^k]\bar{\mathfrak{g}} + (\bar{\mathfrak{g}}^k \cap \bar{\mathfrak{g}}^{k+1}).$$

Proof: Prove (a). Clearly, under the adjoint action of h_0 on $\bar{g}^k$ each vector of $x \in \bar{g}_{c1} \cdots \bar{g}_{c_k}$ satisfies $[h_0,x] = (c_1+\cdots+c_k)x$. Therefore, for any $(c_1,\ldots,c_k) \in \overline{\mathbb{K}}^k$ such that $c_1+\cdots+c_k \neq 0$ the subspace $\bar{g}_{c1}\,\bar{g}_{ck}$ belongs to $[\bar{g}, \bar{g}^k]$. Clearly,

$$\bar{\mathfrak{g}}_{c_1} \cdots \bar{\mathfrak{g}}_{c_{k+1}} \equiv \bar{\mathfrak{g}}_{c_{\sigma(1)}} \cdots \bar{\mathfrak{g}}_{c_{\sigma(k)}} \bar{\mathfrak{g}}_{c_{\sigma(k+1)}} \quad \text{mod } \bar{\mathfrak{g}}^k \cap \bar{\mathfrak{g}}^{k+1}$$

For any permutation $\sigma \in S_{k+1}$.

It is also easy to see that for any $c = (c_1,\ldots,c_{k+1}) \in \mathbb{K}^{k+1} \setminus \{0\}$ there exists a permutation $\sigma \in S_{k+1}$ such that $c_{\sigma(1)} +\cdots+ c_{\sigma(k)} \neq 0$ and, therefore,

$$\bar{\mathfrak{g}}_{c_1} \cdots \bar{\mathfrak{g}}_{c_{k+1}} \subset (\bar{\mathfrak{g}}_{c_{\sigma(1)}} \cdots \bar{\mathfrak{g}}_{c_{\sigma(k)}})\bar{\mathfrak{g}}_{c_{\sigma(k+1)}} + \bar{\mathfrak{g}}^k \cap \bar{\mathfrak{g}}^{k+1} \in [\bar{\mathfrak{g}}, \bar{\mathfrak{g}}^k]\bar{\mathfrak{g}} + (\bar{\mathfrak{g}}^k \cap \bar{\mathfrak{g}}^{k+1})$$

This proves (a).

Prove (b) now. There exists $c \in \overline{\mathbb{K}} \setminus \{0\}$ such that $\bar{\mathfrak{h}}_c = [\bar{g}_c, \bar{g}_{-c}] \neq 0$. Then we obtain the following ongruence:

$$[\bar{\mathfrak{g}}_{c}, \bar{\mathfrak{g}}_0^{k-1}\bar{\mathfrak{g}}_{-c}]\mathfrak{g}_0 \equiv \bar{\mathfrak{g}}_0^{k-1}\bar{\mathfrak{h}}_c\bar{\mathfrak{g}}_0 \mod [\bar{\mathfrak{g}}_{c}, \bar{\mathfrak{g}}_0^{k-1}]\bar{\mathfrak{g}}_{-c}\bar{\mathfrak{g}}_0.$$

Taking into account that $[\bar{g}_{c'}, \bar{g}_0] \subset \bar{g}_{c'}$, we obtain:

$$[\bar{\mathfrak{g}}_{c}, \bar{\mathfrak{g}}_0^{k-1}]\bar{\mathfrak{g}}_{-c}\bar{\mathfrak{g}}_0 \subset \sum_{i=1}^{k-1} \bar{\mathfrak{g}}_0^{i-1}\bar{\mathfrak{g}}_c\bar{\mathfrak{g}}_0^{k-1-i}\bar{\mathfrak{g}}_{-c}\bar{\mathfrak{g}}_0 \subset [\bar{\mathfrak{g}}, \bar{\mathfrak{g}}^k]\bar{\mathfrak{g}} + (\bar{\mathfrak{g}}^k \cap \bar{\mathfrak{g}}^{k+1})$$

by the already proved part (a). Therefore, $\bar{g}_0^{k-1}\,\bar{\mathfrak{h}}_c\bar{g}_0 \subset [\bar{g}, \bar{g}^k]\,\bar{g} + (\bar{g}^k \cap \bar{g}^{k+1})$. Since g is strongly graded, the subspaces $\mathfrak{h}_{c'}$, $c \in \overline{\mathbb{K}} \setminus \{0\}$ span $\bar{g}_{0'}$ and therefore, $\bar{g}_0^{k-1} \subset [\bar{g}, \bar{g}^k]\,\bar{g} + (\bar{g}^k \cap \bar{g}^{k+1})$. This proves (b).

The lemma is proved.

Lemma 4.6: Let g be a semisimple Lie algebra over k. Then for any compatible pair (g,A) one has the following decomposition of the g-module g^k, $k \geq 2$:

$$\mathfrak{g}^k = [\mathfrak{g}, \mathfrak{g}^k] + Z_k(\mathfrak{g}), \qquad [\mathfrak{g}, \mathfrak{g}^k] \cap Z_k(\mathfrak{g}) = \{0\},$$

Where $Z_k(g) = Z(\langle g \rangle)) \cap g^k$, and $Z(\langle g \rangle)$is the center of $\langle g \rangle = \Sigma_{k\geq 0}\, g^k$.

Proof: Clearly, g^k is a semisimple finite-dimensional g-module (under the adjoint action). Therefore, it uniquely decomposes into isotypic components one of which, the component of invariants, is $Z_k(g)$. Denote the sum of all noninvariant isotypic components by $(g^k)+$. By definition, $g^k = (g^k)+ + Z_k(g)$ and $(g^k)+ \cap Z_k(g) = 0$. It remains to prove that $(g^k)+ = [g, g^k]$. Indeed, $[g, g^k] \subseteq (g^k)+$. On the other hand, each nontrivial irreducible g-submodule $V \subset g^k$ is faithful, i.e., $[g,V] = V$ (since $[g,V]$ is always a g-submodule of V). Therefore, $[g, g^k]$ contains all noninvariant isotypic components, i.e., $[g, g^k] \subset (g^k)+$. The double inclusion obtained implies that $(g^k)+ = [g, g^k]$. The lemma is proved.

Now we are ready to prove the theorem part-by-part.

Prove (a): In view of Theorem 3.2(c), it suffices to prove that $\widetilde{(g,A)}$ (F) ⊂ (g,A)(F), that is,

$$I_k(\mathcal{F})\cdot[\mathfrak{g},\mathfrak{g}^{k+1}]\subset(\mathfrak{g},A)(\mathcal{F}),\qquad [\mathcal{F},I_k(\mathcal{F})]\cdot\mathfrak{g}^{k+2}\subset(\mathfrak{g},A)(\mathcal{F}) \tag{4.6}$$

For k ≥0.

We will prove (4.6) by induction on k. First, verify the base of induction at k = 0. Obviously, I_0(F) · [g, g] ⊂ FF · [g, g] ⊂ (g,A)(F). Furthermore, Proposition 4.4(b) taken with k = 1, J = F implies that [F,F] · g^2 ⊂ (g,A)(F).

Now assume that k > 0. Using a part of the inductive hypothesis in the form [F,I_{k-1}(F)] · g^{k+1} ⊂ (g,A)(F) and applying Proposition 4.4(b) with J = [F,I_{k-1}(F)], we obtain (FJ +J)· [g, g^{k+1}] ⊂ (g,A)(F). In its turn, Lemma 1.2(c) taken with $\ell = \infty$ implies that FJ +J = I_k(F). Therefore, we obtain

$$I_k(\mathcal{F})\cdot[\mathfrak{g},\mathfrak{g}^{k+1}]\subset(\mathfrak{g},A)(\mathcal{F}),$$

Which is the first inclusion of (4.6). To prove the second inclusion (4.6), we will use Proposition 4.4(a) with I = I_k(F):

$$(\mathfrak{g},A)(\mathcal{F})\supset[\mathcal{F},I_k(\mathcal{F})]\cdot[\mathfrak{g},\mathfrak{g}^{k+1}]\mathfrak{g}.$$

On the other hand, using the perfectness of the pair (g,A), we obtain:

$$[\mathcal{F},I_k(\mathcal{F})]\cdot[\mathfrak{g},\mathfrak{g}^{k+1}]\mathfrak{g}\equiv[\mathcal{F},I_k(\mathcal{F})]\cdot\mathfrak{g}^{k+2}\mod[\mathcal{F},I_k(\mathcal{F})]\cdot\mathfrak{g}^{k+1}\cup\mathfrak{g}^{k+2}.$$

But Lemma 1.2(a) taken with $\ell = \infty$ implies that I_k(F) ⊂ I_k−1(F), therefore,

$$[\mathcal{F},I_k(\mathcal{F})]\cdot(\mathfrak{g}^{k+1}\cap\mathfrak{g}^{k+2})\subset[\mathcal{F},I_k(\mathcal{F})]\cdot\mathfrak{g}^{k+1}\subset[\mathcal{F},I_{k-1}(\mathcal{F})]\cdot\mathfrak{g}^{k+1}\subset(\mathfrak{g},A)(\mathcal{F})$$

By the inductive hypothesis. This gives the second inclusion of (4.6). Therefore, Theorem 4.3(a) is proved.

Prove (b) now. Lemma 4.5 guarantees that for any strongly graded Lie algebra g one has:

$$[\bar{\mathfrak{g}}, \bar{\mathfrak{g}}^k]\bar{\mathfrak{g}} + (\bar{\mathfrak{g}}^k \cap \bar{\mathfrak{g}}^{k+1}) = \bar{\mathfrak{g}}^{k+1}$$

For all k ≥2, where $\bar{g} = \overline{\mathbb{K}} \otimes g$ is the "algebraic closure" of g. Since the "algebraic closure" commutes with the multiplication and the commutator bracket in A, the restriction of the above equation to $g^{k+1} \subset \bar{g}^{k+1}$ becomes (4.1).

This finishes the proof of Theorem 4.3(b).

Prove (c) now. Since for each semisimple Lie algebra g the compatible pair (g,A) is perfect by the already proved Theorem 4.3(b), Theorem 4.3(a) implies that (g,A)(F) = $\widetilde{(g,A)}$(F). Therefore, in order to finish the proof of Theorem 4.3(c), it suffices to show that

$$\widetilde{(\mathfrak{g}, A)}(\mathcal{F}) = \mathcal{F} \cdot \mathfrak{g} + \sum_{k \geqslant 2} I_{k-1}(\mathcal{F}) \cdot (\mathfrak{g}^k)_+ + [\mathcal{F}, I_{k-2}(\mathcal{F})] \cdot Z_k(\mathfrak{g}).$$

Using Lemma 4.6 and the definition (3.1) of $\widetilde{(g,A)}$ (F), we obtain

$$\begin{aligned}\widetilde{(\mathfrak{g}, A)}(\mathcal{F}) &= \mathcal{F} \cdot \mathfrak{g} + \sum_{k \geqslant 1} I_k(\mathcal{F}) \cdot [\mathfrak{g}, \mathfrak{g}^{k+1}] + [\mathcal{F}, I_{k-1}(\mathcal{F})] \cdot \mathfrak{g}^{k+1} \\ &= \mathcal{F} \cdot \mathfrak{g} + \sum_{k \geqslant 1} (I_k(\mathcal{F}) + [\mathcal{F}, I_{k-1}(\mathcal{F})]) \cdot [\mathfrak{g}, \mathfrak{g}^{k+1}] + [\mathcal{F}, I_{k-1}(\mathcal{F})] \cdot Z_{k+1}(\mathfrak{g}),\end{aligned}$$

Which, after taking into account that $[F, I_{k-1}(F)] \subset I_k(F)$ (and shifting the index of summation), becomes the right-hand side of (4.7). This finishes the proof of Theorem 4.3(b).

Therefore, Theorem 4.3 is proved.

The following is a direct corollary of Theorem 4.3.

Corollary 4.7: Assume that a compatible pair (g,A) is perfect and F is a k-algebra satisfying $I_1(F) = F$. Then

$$(\mathfrak{g}, A)(\mathcal{F}) = \mathcal{F} \cdot \mathfrak{g} + \mathcal{F} \cdot \left[\mathfrak{g}, \langle \mathfrak{g} \rangle\right] + [\mathcal{F}, \mathcal{F}] \cdot \langle \mathfrak{g} \rangle \tag{4.8}$$

$\left(\text{where} \langle g \rangle = \Sigma_{k\geq 1} g^k\right)$.

Proof: First, show by induction that $I_k(F) = F$ for all $k \geq 1$. It follows immediately from Lemma 1.2(c) implying that $I_{k+1}(F) = F[F,I_k(F)]+[F,I_k(F)]$.

This and (3.1) imply that

$$\widetilde{(\mathfrak{g}, A)}(\mathcal{F}) = \mathcal{F} \cdot \mathfrak{g} + \sum_{k \geqslant 1} \mathcal{F} \cdot \left[\mathfrak{g}, \mathfrak{g}^{k+1}\right] + [\mathcal{F}, \mathcal{F}] \cdot \mathfrak{g}^{k+1} = \mathcal{F} \cdot \mathfrak{g} + \mathcal{F} \cdot \left[\mathfrak{g}, \langle \mathfrak{g} \rangle\right] + [\mathcal{F}, \mathcal{F}] \cdot \langle \mathfrak{g} \rangle.$$

This and Theorem 4.3 finish the proof.

Remark 4.8: The condition $I_1(F) = F$ holds for each noncommutative simple unital algebra F, e.g., for each noncommutative skew-field F containing $\mathbb{K}$. Therefore, for all such algebras and any perfect pair (g,A), the Lie algebra (g,A)(F) is given by the relatively simple formula (4.8), which also complements (2.6).

The following result is a specialization of Theorem 4.3 to the case when $g = sl_2(\mathbb{K})$.

Theorem 4.9: Let A an object of Alg_1 containing $sl_2(\mathbb{K})$ as a Lie subalgebra. Then

$$(sl_2(\Bbbk), A)(\mathcal{F}) = \mathcal{F} \cdot sl_2(\Bbbk) + \left[\mathcal{F} \cdot 1, Z_1(A, \mathcal{F})\right] + \sum_{i \geqslant 1} Z_i(A, \mathcal{F}) \cdot V_{2i}, \tag{4.9}$$

where

$$Z_i(A, \mathcal{F}) = \sum_{j \geqslant 0} I_{i+2j-1}(\mathcal{F}) \cdot \Delta^j,$$

$\Delta = 2EF + 2FE + H^2$ is the Casimir element, and V_{2i} is the $sl_2(\mathbb{K})$-submodule of A generated by E^i . In particular, if A = End (V), where V is a simple (m + 1)-dimensional $sl_2(\mathbb{K})$-module, then

$$(sl_2(\Bbbk), A)(\mathcal{F}) = [\mathcal{F}, \mathcal{F}] \cdot 1 + \sum_{k=1}^{m} I_{k-1}(\mathcal{F}) \cdot V_{2k}. \tag{4.10}$$

Proof: Prove (4.9). Clearly, each g^k is a finite-dimensional $sl_2(\mathbb{K})$-module generated by the highest weight vectors $\Delta^j E^i$, $i,j \geq 0$, $i + 2j \leq k$. That is, in notation of (4.3), one has

$$(\mathfrak{g}^k)_+ = \sum_{i>0, j\geqslant 0, i+2j\leqslant k} \Delta^j \cdot V_{2i},$$

Where the sum is direct (but some summands may be zero) and

$$V_{2i} = \sum_{r=-i}^{i} \Bbbk \cdot (ad\ F)^{i+r}(E^i)$$

Is the corresponding simple $sl_2(\mathbb{K})$-module; and

$$Z_k(\mathfrak{g}) = \sum_{1\leqslant j\leqslant k/2} \Bbbk \cdot \Delta^j$$

Where the sum is direct. Therefore, taking into account that $I_k(F) \subset I_k-1(F)$, the equation (4.3) simplifies to

$$\begin{aligned}(sl_2(\Bbbk), A)(\mathcal{F}) &= \mathcal{F} \cdot \mathfrak{g} + \sum_{i>0, j\geqslant 0} I_{i+2j-1}(\mathcal{F}) \cdot \Delta^j V_{2i} + \sum_{j\geqslant 1} [\mathcal{F}, I_{2j-2}(\mathcal{F})] \cdot \Delta^j \\ &= \mathcal{F} \cdot sl_2(\Bbbk) + [\mathcal{F} \cdot 1, Z_1(A, \mathcal{F})] + \sum_{i\geqslant 1} Z_i(A, \mathcal{F}) \cdot V_{2i}.\end{aligned}$$

This finishes the proof of (4.9).

Prove (4.10). Indeed, now $\Delta \in k \setminus \{0\}$, $E^k = 0$ for k>m. Therefore

$$Z_k(A,\mathcal{F}) = \sum_{j \geqslant 0} I_{k+2j-1}(\mathcal{F}) \cdot \Delta^j = \sum_{j \geqslant 0} I_{k+2j-1}(\mathcal{F}) \cdot 1 = I_{k-1}(\mathcal{F}) \cdot 1$$

because $I_k(F) \subset I_k - 1(F)$. Finally, using (4.9), we obtain:

$$\begin{aligned}(sl_2(\Bbbk), A)(\mathcal{F}) &= \mathcal{F} \cdot sl_2(\Bbbk) + [\mathcal{F}, Z_1(A,\mathcal{F})] \cdot 1 + \sum_{i \geqslant 1} Z_i(A,\mathcal{F}) \cdot V_{2i} \\ &= [\mathcal{F},\mathcal{F}] \cdot 1 + \sum_{1 \leqslant k \leqslant m} I_{k-1}(\mathcal{F}) \cdot V_{2k}.\end{aligned}$$

Theorem 4.9 is proved.

N –GROUPS

Throughout the section we assume that each object of N is a unital k-algebra, i.e., N is a sub-category of Alg_1.

From N -Lie Algebras to N -groups and Generalized K1-theories

In this section we use F-algebras and the category $LieAlg_N$ defined in Section 2.

Definition 5.1: An affine N -group is a triple (F,G,A), where F is an object of N , A is an F-algebra in Alg1 (i.e., $\imath$:F → A respects the unit), and G is a subgroup of the group of units $A^\times$ such that G contains the image $\imath(F^\times) = \imath(F)^\times$. A morphism $(F_1,G_1,A_1) \to (F_2,G_2,A_2)$ of affine N -groups is a pair (ϕ,ψ), where $\phi : F_1 \to F_2$ is a morphism in N and $\psi : G_1 \to G_2$ is a group homomorphism such that $\psi \circ \imath_1 | F_1^\times = \imath_2 | F_1^\times 0\psi$.

Denote by $Gr_{\mathcal{N}}$ the category of affine $\mathcal{N}$ -groups.

Let $LieAlg_{\mathcal{N};1}$ be the sub-category of $LieAlg_{\mathcal{N}}$ whose objects are triples (F,L,A), where F is an object of Alg_1, A is an F-algebra in Alg_1, and L is a Lie subalgebra of A invariant under the adjoint action of $\imath(F)$; morphisms in $LieAlg_{\mathcal{N};1}$ are those morphisms in $LieAlg_{\mathcal{N}}$ which respect the unit.

Given an object $(F, \mathcal{L}, A)$ of $\mathrm{LieAlg}_{\mathcal{N};1}$, define the triple $\mathrm{Exp}(F, \mathcal{L}, A) := (F,G,A)$, where G is the subgroup of $A^\times$ generated by $\imath(F)^\times$ and by the stabilizer $\{g \in A^\times: g\mathcal{L}g^{-1} = \mathcal{L}\}$ of $\mathcal{L}$ in $A^\times$.

Given an object (F,G,A) of $\mathrm{Gr}_{\mathcal{N}}$, define the triple $\mathrm{Lie}(F,G,A) := (F, \mathcal{L}, A)$, where $\mathcal{L}$ is the Lie subalgebra of A generated (over $\mathbb{K}$) by the set $\{g \cdot \imath(f) \cdot g^{-1}: g \in G, f \in F\}$, that is, $\mathcal{L}$ is the smallest Lie subalgebra of A containing $\imath(F)$ and invariant under conjugation by G (therefore, $\mathcal{L}$ is invariant under the adjoint action of the subalgebra $\imath(F)$).

The following result is obvious.

Lemma 5.2: For each object $(F, \mathcal{L}, A)$ of $\mathrm{LieAlg}_{\mathcal{N};1}$ the triple $\mathrm{Exp}(F, \mathcal{L}, A)$ is an affine $\mathcal{N}$-group; and for any affine $\mathcal{N}$-group (F,G,A) the triple $\mathrm{Lie}(F,G,A)$ is an object of $\mathrm{LieAlg}_{\mathcal{N};1}$.

Remark 5.3: The operations Exp and Lie are analogues of the Lie correspondence (between Lie algebras and Lie groups). However, similarly to the operations L_i from Lemma 2.2, in general they are not functors.

Composing these two operations with each other and the operations L_i, we can obtain a number of affine $\mathcal{N}$-groups out of a given $\mathcal{N}$-Lie algebra and vice versa.

By definition, one has a natural (forgetful) projection functor $\pi : \mathrm{Gr}_{\mathcal{N}} \to \mathcal{N}$ by $\pi(F,G,A) = F$ and $\pi(\phi,\psi) = \phi|F_1$.

A noncommutative current group (or simply $\mathcal{N}$-current group) is any functor $\mathfrak{G}: \mathcal{N} \to \mathrm{Gr}_{\mathcal{N}}$ such that $\pi \circ \mathfrak{G} = \mathrm{Id}_{\mathcal{N}}$ (i.e., $\mathfrak{G}$ is a section of π).

Note that if $\mathcal{N} = (F, \mathrm{Id}_F)$ has only one object F and only the identity arrow Id_F, then the $\mathcal{N}$-current group is simply any object of $\mathrm{Gr}_{\mathrm{Alg}1}$ of the form (F,G,A). In this case, we will sometimes refer to G an F-current group.

The above arguments allow for constructing a number of F-current groups out of F-current Lie algebras and vice versa. In a different situation, for any subcategory N of Alg_1, we will construct below a class of $\mathcal{N}$ -current groups associated with compatible pairs (g,A). More general $\mathcal{N}$ -current groups will be considered elsewhere.

Similarly to Section 2, given an object F of Alg_1 and a group G, we refer to a group homomorphism $\iota : F^{\times} \to G$ in Alg as an F-group structure on G (we will also refer to G an F-group).

Definition 5.4: A decorated group is a pair (F,G), where F is an object of Alg_1 and G is an F-group.

We denote by DecGr the category whose objects are decorated groups and morphisms are pairs $(\phi,\psi):(F_1,G_1) \to (F_2,G_2)$, where $\phi : F_1 \to F_2$ is a morphism in Alg_1 and $\psi : G_1 \to G_2$ is a group homomorphism such that $\psi \circ \iota_1 = \iota_2 \circ \phi|\, F_1^{\times}$.

In particular, one has a natural (forgetful) projection functor π :DecGr $\to Alg_1$ by $\pi(F,G) = F$ and $\pi(\phi,\psi) = \phi$. Note also that for any object (F,G,A) of $Gr_{\mathcal{N}}$ the pair (F,G) is a decorated group, therefore, the projection $(F,G,A) \mapsto (F,G)$ defines a (forgetful) functor $Gr_{\mathcal{N}} \to$ DecGr.

Definition 5.5: A generalized K_1-theory is a functor K: $\mathcal{N} \to$ DecGr such that $\pi \circ K = \pi|_{\mathcal{N}}$.

In what follows we will construct a number of generalized K_{1-}theories as compositions of an $\mathcal{N}$ -current group s: $\mathcal{N} \to Gr_{\mathcal{N}}$ with a certain functors κ from $Gr_{\mathcal{N}}$ to the category of decorated groups.

Given an object (F,G,A) of $Gr_{\mathcal{N}}$, we define ${}_{\kappa com}(F,G,A) := G/[G,G]$, where [G,G] is the (normal) commutator subgroup of G.

Lemma 5.6: The correspondence $(F,G,A) \mapsto {}_{\kappa com}(F,G,A)$ defines a functor ${}_{\kappa com}$ from $Gr_{\mathcal{N}}$ to the category of decorated abelian groups. In particular, for any $\mathcal{N}$ -current group $\mathfrak{G}$: $\mathcal{N} \to Gr_{\mathcal{N}}$ the composition ${}_{\kappa com} \circ \mathfrak{G}$ is a generalized K_{1-}theory.

Note that each K_1-theory defined by Lemma 5.6 is still commutative. Below we will construct a number of noncommutative nilpotent K_1-theories in a similar manner.

Definition 5.7: Let A be an F-algebra. Recall that a subset S of A is called a nilpotent if $S^n = 0$ for some n. In particular, an element $e \in A$ is nilpotent if {e} is nilpotent, i.e., $e^n = 0$. We denote by A_{nil} the set of all nilpotent elements in A. Note that A_{nil} is invariant under adjoint action of the group of units $A^\times$. We say that an element $e \in A$ is an F-stable nilpotent if $(F \cdot e \cdot F)^n = 0$ for some $n > 0$. Denote by $(F,A)_{nil}$ the set of all F-stable nilpotent element of A. Denote also by $A_{nil}^{F^\times}$ nil and $(F,A)_{nil}^{F^\times}$ the centralizers of $F^\times$ in A_{nil} and $(F,A)_{nil}$ respectively.

In particular, taking A = F, we see that $f \in (F,F)_{nil}$ if and only if the ideal $FfF \subset F$ is nilpotent. Note also that if the image $\imath(F)$ is in the center of A, then $(F,A)_{nil} = A_{nil}$.

Let (F,G,A) be an object of $\mathrm{Gr}_{\mathcal{N}}$ and let S be a subset of A_{nil}, denote by $E_S = E_S(F,G,A)$ the subgroup of G generated by all $1 + xsx^{-1}$, $s \in S$, $x \in G$. Clearly, E_S is a normal subgroup of G. Then denote the quotient group G/E_S by:

$$\begin{cases} \kappa_{nil}(\mathcal{F},\mathcal{G},\mathcal{A}) & \text{if } S = \mathcal{A}_{nil}, \\ \kappa_{stnil}(\mathcal{F},\mathcal{G},\mathcal{A}) & \text{if } S = (\mathcal{F},\mathcal{A})_{nil}, \\ \kappa_{nil,inv}(\mathcal{F},\mathcal{G},\mathcal{A}) & \text{if } S = \mathcal{A}_{nil}^{\mathcal{F}^\times}, \\ \kappa_{stnil,inv}(\mathcal{F},\mathcal{G},\mathcal{A}) & \text{if } S = (\mathcal{F},\mathcal{A})_{nil}^{\mathcal{F}^\times}. \end{cases} \quad (5.1)$$

Lemma 5.8: Each of the four correspondences

$$(\mathcal{F},\mathcal{L},\mathcal{A}) \mapsto \kappa(\mathcal{F},\mathcal{L},\mathcal{A}),$$

Where $\kappa = \kappa_{nil}, \kappa_{stnil}, \kappa_{nil,inv}, \kappa_{stnil,inv}$, defines a functor

$$\kappa : \mathbf{Gr}_{\mathcal{N}} \to \mathbf{DecGr}.$$

In particular, for any $\mathcal{N}$ -current group $\mathfrak{G}: \mathcal{N} \to \mathrm{Gr}_{\mathcal{N}}$ the composition $\kappa \circ \mathfrak{G}$ is a generalized K_1-theory $\mathcal{N} \to \mathrm{DecGr}$.

Proof: Clearly, in each case of (5.1), the association $A \mapsto S = S_A$ is functorial, i.e., commutes with morphisms of F-algebras. Therefore, the association $A \mapsto E_S$ is also functorial in all four cases (5.1). This finishes the proof of the lemma.

We will elaborate examples of generalized (noncommutative) K_1-groups in a separate paper.

N -current Groups for Compatible Pairs

Here we keep the notation of Section 2. Since both F and A are now unital algebras, so is $F \otimes A = F \cdot A$.

Note that for the N -Lie algebra $s : F \mapsto (F,(g,A)(F),F \cdot A)$ the corresponding $\mathcal{N}$ -current group is of the form $(F,G_{g,A}(F),F\,A)$, where $G_{g,A}(F)$ is the normalizer of $(g,A)(F)$ in $(F\,A)^\times$ (i.e., $G_{g,A}(F) = \{g \in (F\,A)^\times : g\,(g,A)(F)\,g^{-1} = (g,A)(F)\}$).

The following facts are obvious.

Lemma 5.9: Let (g,A) be a compatible pair and $S \subset g$ be a generating set of g as a Lie algebra. Then for any object F of Alg_1 an element $g \in (F\,A)^\times$ belongs to $G_{g,A}(F)$ if and only if:

$$g(u \cdot x)g^{-1} \subset (\mathfrak{g}, A)(\mathcal{F}) \tag{5.2}$$

For all $x \in S$, $u \in F$.

Lemma 5.10: For each compatible pair of the form $(g,A) = (sl_n(\mathbb{K}),M_n(\mathbb{K}))$ and an object F of $\mathcal{N}_1$ one has: $G_{g,A}(F) = GL_n(F) = (F\,A)^\times$.

In what follows we will consider compatible pairs of the form (g,End(V)) where V is a simple finite-dimensional g-module. By choosing an appropriate basis in V , we identify A = End(V) with $M_n(\mathbb{K})$ so that $G_{g,A}(F) \subset GL_n(F)$. In all cases to be considered, we will compute the "Cartan subgroup" $(F^\times)^n \cap G_{g,A}(F)$ of $G_{g,A}(F)$.

Let Φ_0 be the bilinear form on the $\mathbb{K}$-vector space $V = \mathbb{K}^n$ given by:

$$\Phi_0(x, y) = x_1 y_n + x_2 y_{n-1} + \cdots + x_n y_1.$$

Also define the bilinear form Φ_1 on $\mathbb{K}^{2m}$ by:

$$\Phi_1(x, y) = x_1 y_{2m} + x_2 y_{2m-1} + \cdots + x_m y_m - x_{m+1} y_{m-1} \cdots - x_{2m} y_1.$$

Proposition 5.11: Let F be an object of Alg_1, $A = M_n(\mathbb{K})$, and suppose that either $g = o(\Phi_0)$ or $g = o(\Phi_1)$ and $n = 2m$. Then an invertible diagonal matrix $D = diag(f_1,\ldots,f_n) \in GL_n(F)$ belongs to $G_{g,A}(F)$ if and only if

$$f_i f_{n-i+1} - f_1 f_n \in I_1(\mathcal{F}) = \mathcal{F}[\mathcal{F}, \mathcal{F}]$$

For i = 1,...,n.

Proof: We will prove the proposition for $g = o(\Phi_0)$ (the proof for $g = o(\Phi_1)$ is nearly identical). It is easy to see that g is a Lie subalgebra of $sl_n(\mathbb{K})$ generated by all $e_{ij} := E_{ij} - E_{n-j+1,n-i+1,}$ i,j = 1,...,n. therefore, Lemma 5.9 (with $S = \{e_{ij}\}$) guarantees that $D = (f_1,\ldots,f_n) \in (F^{\times})^n$ belongs to $G_{g,A}(F)$ if and only if $D(u \cdot e_{ij})D^{-1} \subset (g,A)(F)$ for all $u \in F$, i,j = 1,...,n. Note

$$D(u \cdot e_{ij})D^{-1} = f_i u f_j^{-1} E_{ij} - f_{j'} u f_{i'}^{-1} E_{j',i'} = f_i u f_j^{-1} e_{ij} + \delta_{ij}(u) E_{j',i'},$$

where $\delta_{ij}(u) = f_i u f_j^{-1} - f_j u f_i^{-1}$ i and $I' = n + 1, - i$, $j' = n + 1 - j$. Therefore, taking into account that $(g,A)(F) = [F,F] \cdot 1 + F \cdot g + I_1(F) \cdot sl_n(\mathbb{K})$ by Corollary 2.15, we see that $D(u \cdot e_{ij})^{D-1} \in (g,A)(F)$ if and only if $\delta_{ij}(u) \in I_1(F)$. Note that $\delta_{ij}(u) \equiv u\delta_{ij}(1) \mod I_1(F)$. Since $I_1(F)$ is an ideal, $u\delta_{ij}(1)$

$\in I_1(F)$ for all $u \in F$ if and only if $\delta_{ij}(1) \in I_1(F)$, i.e., $fi f_j^{-1} - f_j f_i^{-1} \mod I_1(F)$.

Taking into account that $f_i f_j^{-1} - f_j f_i^{-1} \equiv (f_i f_i - f_j f_j) f_j^{-1} f_i^{-1} \mod I_1(F)$, we see that $D \in G_{g,A}(F)$ if and only if $f_i f_{n+1-i} - f_j f_{n+1-j} \in I_1(F)$ for all i,j . Clearly, it suffices to take j = 1. The proposition is proved.

To formulate the main result of this section we need the following notation. For any $\ell \geq 0$ and any $m_1,...,m_{\ell+1} \in F$ denote

$$\Delta^{(\ell)}(m_1,\ldots,m_{\ell+1}) = \sum_{k=0}^{\ell}(-1)^k\binom{\ell}{k}m_{k+1} \tag{5.3}$$

And refer to it as the th difference derivative. Clearly,

$$\Delta^{(\ell)}(m_1,\ldots,m_{\ell+1}) = \Delta^{(\ell-1)}(m_1,\ldots,m_\ell) - \Delta^{(\ell-1)}(m_2,\ldots,m_{\ell+1}).$$

Let $A_n = M_n(\mathbb{K}) = \mathrm{End}(V_{n-1})$, where V_{n-1} is the n-dimensional irreducible $sl_2(\mathbb{K})$-module. Then $G_{sl2(\mathbb{K})}, A_n(F)$ is naturally a subgroup of $GL_n(F)$.

Main Theorem 5.12: For any object F of Alg_1, the "Cartan subgroup" $(F^\times)^n \cap G_{sl2(\mathbb{K})}, A_n(F)$ consists of all $D = (f_1,...,f_n) \in (F^\times)^n$ such that:

$$\Delta^{(k)}(f_1f_2^{-1},\ldots,f_{k+1}f_{k+2}^{-1}) \in I_k(\mathcal{F}) \tag{5.4}$$

for k = 1,...,n − 2.

Proof: We will prove the theorem in several steps. First, we prove Proposition 5.13 by using Lemmas 5.14 and 5.15. Then we prove Lemma 5.16 and Proposition 5.17. The proof of the Proposition is based on Lemma 5.18. The final step in the proof of Theorem 5.12 is Theorem 5.19. This theorem required Proposition 5.21, Lemma 5.23, and Proposition 5.24. Our proofs of Propositions 5.21 and 5.24 use Lemmas 5.22 and 5.25 correspondingly.

We start with a characterization of $(F^\times)^n \cap G_{sl2(\mathbb{K})}, A_n(F)$.

Proposition 5.13: A diagonal matrix $D = (f_1,...,f_n) \in (F^\times)^n$ belongs to the group $G_{sl2(\mathbb{K})}, A_n(F)$ if and only if:

$$\Delta^{(k)}\left(f_1uf_2^{-1}, \ldots, f_{k+1}uf_{k+2}^{-1}\right) \in I_k(\mathcal{F}), \tag{5.5}$$

$$\Delta^{(k)}\left(f_nuf_{n-1}^{-1}, \ldots, f_{n-k}uf_{n-1-k}^{-1}\right) \in I_k(\mathcal{F}). \tag{5.6}$$

for k = 1,...,n − 2 and all u ∈ F

Proof: Denote by $(A_n)_k$ the set of all x ∈ A_n such that [H,x] = kx. Clearly, $(A_n)_k \neq 0$ if and only if k is even and −2(n − 1) ≤ k ≤ 2(n − 1). In fact, $(A_n)_{2k}$ is the span of all those E_{ij} such that 2(j − i) = k. In particular, E ∈ $(A_n)_2$. Denote also $(sl_2(\Bbbk),A_n)(F)_k := (sl_2(\Bbbk),A_n)(F) \cap F \cdot (A_n)k$.

Lemma 5.14: The components $(sl_2(\Bbbk),A_n)(F)_j$, j = −2, 2 are given by:

$$\left(sl_2(\Bbbk), A_n\right)(\mathcal{F})_2 = \sum_{k=0}^{n-2} I_k(\mathcal{F}) \cdot E^{(k)}, \qquad \left(sl_2(\Bbbk), A_n\right)(\mathcal{F})_{-2} = \sum_{k=0}^{n-2} I_k(\mathcal{F}) \cdot F^{(k)}, \tag{5.7}$$

Where

$$E^{(k)} = \sum_{i=k+1}^{n-1} i\binom{i-1}{k} E_{i,i+1}, \qquad F^{(k)} = \sum_{i=k+1}^{n-1} i\binom{i-1}{k} E_{n+1-i,n-i}$$

for k = 0, 1,...,n − 2 form a basis for $(A_n)_2$ and $(A_n)_{-2}$ respectively.

Proof: Let us prove the formula for j = 2. Let p0(x),...,p_n−2(x) be any polynomials in $\Bbbk$ [x] such that degp_k(x) = k for all k. Then it is easy to see that.

$$\left(sl_2(\Bbbk), A_n\right)(\mathcal{F})_2 = \sum_{k=0}^{n-2} I_k(\mathcal{F}) \cdot \left(p_k(H) \cdot E\right).$$

Take

$$p_k(H) = \binom{H'}{k}$$

Where

$$H' = \tfrac{1}{2}(n \cdot 1 - H) = \sum_{i=1}^{n} (i-1) E_{ii}$$

Then

$$p_k(H) = \sum_{i=1}^{n} \binom{i-1}{k} E_{ii}$$

and

$$p_k(H) \cdot E = \left(\sum_{i=1}^{n} \binom{i-1}{k} E_{ii} \right) \left(\sum_{i=1}^{n} i E_{i,i+1} \right) = E^{(k)}.$$

To prove the formula for j = −2, it suffices to conjugate the formula for j = 2 with the matrix of the longest permutation $w_0 = \sum_{i=1}^{n} E_{i,n+1-i}$ i.e., apply the involution $E_{ij} \mapsto E_{n+1-i,n+1-j}$.

Lemma 5.15: For each diagonal matrix $D = (f_1, f_2, \ldots, f_n) \in (F^\times)^n$ and $u \in F$ one has

$$D(u \cdot E)D^{-1} = \sum_{k=0}^{n-2} \Delta^{(k)}(f_1 u f_2^{-1}, \ldots, f_{k+1} u f_{k+2}^{-1}) \cdot (-1)^k E^{(k)}$$

$$D(u \cdot F)D^{-1} = \sum_{k=0}^{n-2} \Delta^{(k)}(f_n u f_{n-1}^{-1}, \ldots, f_{n-k} u f_{n-1-k}^{-1}) \cdot (-1)^{k+1} F^{(k)},$$

where $\Delta^{(k)}$ is the kth divided difference as in (5.3).

Proof: It is easy to see that the elements $E^{(k)}, F^{(k)}$ satisfy:

$$i E_{i,i+1} = \sum_{k=i-1}^{n-2} (-1)^{k+1-i} \binom{k}{i-1} E^{(k)},$$

$$i E_{n+1-i,n-i} = \sum_{k=i-1}^{n-2} (-1)^{k+1-i} \binom{k}{i-1} F^{(k)} \qquad (5.8)$$

for i = 1,...,n − 1.

Furthermore,

$$D(u\cdot E)D^{-1}=\sum_{i=1}^{n-1} f_i u f_{i+1}^{-1}\cdot iE_{i,i+1}=\sum_{i=1}^{n-1} f_i u f_{i+1}^{-1}\cdot\sum_{k=0}^{n-2}(-1)^{k+1-i}\binom{k}{i-1}E^{(k)}$$

$$=\sum_{k=0}^{n-2}\sum_{i=1}^{n-1}(-1)^i\binom{k}{i-1}f_i u f_{i+1}^{-1}\cdot(-1)^{k+1}E^{(k)}$$

$$=\sum_{k=i-1}^{n-2}\Delta^{(k)}(f_1 u f_2^{-1},\ldots,f_{k+1}uf_{k+2}^{-1})\cdot(-1)^{k+1}E^{(k)}.$$

The formula for $D(u \cdot F)D^{-1}$ follows. The lemma is proved.

Now we are ready to finish the proof of Proposition 5.13..

Since the set $S = \{E,F\}$ generates $sl_2(\Bbbk)$, Lemma 5.9 guarantees that $D \in GL_n(F)$ belongs to $G_{sl_2(\Bbbk),A_n}(F)$ if and only if $D(u \cdot E)D^{-1}, D(u \cdot F)D^{-1} \in (sl_2(k),A_n)(F)$ for all $u \in F$. Using the obvious fact that $D(u \cdot E)D^{-1} \subset F \cdot (A_n)_2$ for all $D \in (F^\times)^n$, $u \in F$, we see that $D(u \cdot E)D^{-1} \in (sl_2(\Bbbk),An)(F)$ if and only if $D(u \cdot E)D^{-1} \in (sl_2(\Bbbk),A_n)(F)_2$. In turn, using Lemmas 5.14 and 5.15, we see that this is equivalent to

$$D(u\cdot E)D^{-1}=\sum_{k=0}^{n-2}\Delta^{(k)}(m_1,\ldots,m_{k+1})\cdot(-1)^{k+1}E^{(k)}\in\sum_{k=0}^{n-2}I_k(\mathcal{F})\cdot E^{(k)},$$

which, because the $E^{(0)}$,...,$E^{(n-2)}$ are linearly independent, is equivalent to (5.5). Applying the above argument to $D(u \cdot F)D^{-1}$, we obtain

$$D(u\cdot E)D^{-1}=\sum_{k=0}^{n-2}\Delta^{(k)}(f_n u f_{n-1}^{-1},\ldots,f_{n-k}uf_{n-1-k}^{-1})\cdot(-1)^{k+1}F^{(k)}\in\sum_{k=0}^{n-2}I_k(\mathcal{F})\cdot F^{(k)},$$

which gives (5.6).

The proposition is proved.

Furthermore, we need to establish some basic properties of inclusions (5.4).

Lemma 5.16: Let $m_1, m_2, \ldots, m_\ell$ be elements of F. The following are equivalent:

a. $\Delta^{(k)}(m_1, \ldots, m_{k+1}) \in I_k(\mathcal{F})$ *for all* $1 \leqslant k \leqslant \ell - 1$.

b. $\Delta^{(j-i)}(m_i, \ldots, m_j) \in I_{j-i}(\mathcal{F})$ *for all* $1 \leqslant i \leqslant j \leqslant \ell$.

Proof: The implication (b) ⇒ (a) is obvious. Prove the implication (a) ⇒ (b). Denote

$$m_{ij} := \Delta^{(j-i)}(m_i, \ldots, m_j) \tag{5.9}$$

for all $1 \leq i \leq j \leq \ell$. In particular, $m_{ii} = m_i$ and:

$$m_{ij} = m_{i,j-1} - m_{i+1,j} \tag{5.10}$$

for all $1 \leq i \leq j \leq \ell$.

Prove that the inclusions $m_{1,k+1} \in I_k(F)$ for $0 \leq k \leq \ell$ imply inclusions $m_{ij} \in I_{j-i}(F)$ for all $i \leq j$ such that $j - i \leq \ell$. We proceed by induction on i. The basis of the induction, when i = 1, is obvious. Assume that i > 1 and $j \leq \ell - 1 - i$. Then the inclusions (the inductive hypothesis)

$$m_{i-1,j} = m_{i-1,j-1} - m_{ij} \in I_{j+1-i}(\mathcal{F}) \subset I_{j-i}(\mathcal{F})$$

and $mi-1_{,j-1} \in I_{j-i}(F)$ imply that $m_{ij} \in I_{j-i}(F)$. This finishes the proof of the implication (a) ⇒ (b). The lemma is proved.

Proposition 5.17: Let $m_1, m_2, \ldots, m_\ell$ be invertible elements of F. The following are equivalent:

a. $\Delta^{(j-i)}(m_i, \ldots, m_j) \in I_{j-i}(\mathcal{F})$ *for all* $i \leqslant j$.

b. $\Delta^{(j-i)}(m_i^{-1}, \ldots, m_j^{-1}) \in I_{j-i}(\mathcal{F})$ *for all* $i \leqslant j$

Proof: We need the following notation. Similarly to (5.9) denote

$$m_{ij} := \Delta^{(j-i)}(m_i, \ldots, m_j), \qquad m_{ij}^* := \Delta^{(j-i)}(m_i^{-1}, \ldots, m_j^{-1}) \tag{5.11}$$

for all $1 \leq i \leq j \leq$. In particular, $m_{ii}^* = m_i^{-1}$ and $m_{12}^* = -m_1^{-1}(m_1 - m_2)m_2^{-1}$ $= -m_{11}^* m_{12} m_{22}^*$ We need the following recursive formula for m_{ij}^*.

Lemma 5.18: In the above notation, we have for all $1 \leq i < j \leq \ell$.

$$m_{ij}^* = \sum_{\substack{i \leqslant i_1 \leqslant j_1 \leqslant j, i \leqslant i_2 < j_2 \leqslant j, i \leqslant i_3 \leqslant j_3 \leqslant j \\ j_1 - i_1 + j_2 - i_2 + j_3 - i_3 = j - i}} c_{i,i_1,i_2,i_3}^{j,j_1,j_2,j_3} m_{i_1,j_1}^* m_{i_2,j_2} m_{i_3,j_3}^*, \tag{5.12}$$

where the coefficients are translation-invariant integers:

$$c_{i+1,i_1+1,i_2+1,i_3+1}^{j+1,j_1+1,j_2+1,j_3+1} = c_{i,i_1,i_2,i_3}^{j,j_1,j_2,j_3}.$$

Proof: We proceed by induction on $j - i$. If $j = i + 1$, we obtain:

$$m_{i,i+1}^* = -m_i^{-1}(m_i - m_{i+1})m_{i+1}^{-1} = -m_{ii}^* m_{i,i+1} m_{i+1,i+1}^* = -m_{i+1,i+1}^* m_{i,i+1} m_{i,i}^*.$$

Next, assume that $j - i > 1$. Then, using the translation invariance of the coefficients in (5.12) for $m_{ij\,j-1}^*$, we obtain.

$$m_{ij}^* = m_{i,j-1}^* - m_{i+1,j}^* = \sum_{\substack{i \leqslant i_1 \leqslant j_1 \leqslant j-1, i \leqslant i_2 < j_2 \leqslant j-1, i \leqslant i_3 \leqslant j_3 \leqslant j-1 \\ j_1 - i_1 + j_2 - i_2 + j_3 - i_3 = j-1-i}} c_{i,i_1,i_2,i_3}^{j-1,j_1,j_2,j_3} \delta_{i_1,i_2,i_3}^{j_1,j_2,j_3},$$

Where

$$\delta^{j_1,j_2,j_3}_{i_1,i_2,i_3} = m^*_{i_1,j_1} m_{i_2,j_2} m^*_{i_3,j_3} - m^*_{i_1+1,j_1+1} m_{i_2+1,j_2+1} m^*_{i_3+1,j_3+1}.$$

Furthermore

$$\begin{aligned}\delta^{j_1,j_2,j_3}_{i_1,i_2,i_3} &= m^*_{i_1,j_1+1} m_{i_2,j_2} m^*_{i_3,j_3} + m^*_{i_1+1,j_1+1}\left(m_{i_2,j_2} m^*_{i_3,j_3} - m_{i_2+1,j_2+1} m^*_{i_3+1,j_3+1}\right)\\ &= m^*_{i_1,j_1+1} m_{i_2,j_2} m^*_{i_3,j_3} + m^*_{i_1+1,j_1+1} m_{i_2,j_2+1} m^*_{i_3,j_3} + m^*_{i_1+1,j_1+1} m_{i_2+1,j_2+1} m^*_{i_3,j_3+1}.\end{aligned}$$

This proves the formula (5.12) for m^*_{ij}. The lemma is proved.

We are ready to finish the proof of Proposition 5.17 now.

Due to the symmetry, it suffices to prove only one implication, say (a) ⇒ (b). The desired implication follows inductively from (5.11).

Note that, in the view of Lemma 5.16, the condition (a) (respectively the condition (b)) of Proposition 5.17 for mi = fi f^{-1}_{i+1} i = 1,...,n − 2, is a particular case of (5.5) (respectively of (5.6)) with u = 1.

Furthermore, we need one more result in order to finish the proof of Theorem 5.12.

Theorem 5.19: The inclusions (5.4) imply the inclusions (5.5) (and, therefore, (5.6)).

Proof: To prove the theorem we will develop a formalism of homogeneous maps F → F (relative to the ideals I_k(F)).

Definition 5.20: We say that a $\Bbbk$-linear map ∂ :F → F is homogeneous of degree if ∂(Ik(F)) ⊂ $I_{k+\ell}$(F) for all k≥ 0; denote by End$^{(\ell)}$(F) the set of all such maps.

Lemma 1.4 guarantees that for each $f_1 \in I_{\ell_1}$ (F), $f_2 \in I_{\ell_2}$ (F), the map F → F given by u ↦ $f_1 u f_2$ is homogeneous of degree $\ell_1 + \ell_2$.

We construct a number of homogeneous maps of degree 1 as follows. For an invertible element $m \in F^\times$ define $\partial_m : F \to F$ by

$$\partial_m(u) = mum^{-1} - u = \left[m, um^{-1}\right].$$

Clearly, $\partial_m : F \to F$ is homogeneous of degree 1.

Proposition 5.21: Let $m_1, m_2, \ldots, m_\ell$ be invertible elements of F such that, in the notation (5.9), one has $m_{ij} \in I_{j-i}(F)$ for all $1 \leq i \leq j \leq \ell$. Then

$$\Delta^{(j-i)}(\partial_{m_i}, \ldots, \partial_{m_j}) \in End^{(j+1-i)}(\mathcal{F})$$

for all $1 \leq i \leq j \leq \ell$.

Proof: We need the following notation. Similarly to (5.9), denote

$$\partial_{ij} = \Delta^{(j-i)}(\partial_{m_i}, \ldots, \partial_{m_j}) \tag{5.13}$$

for $1 \leq i \leq j \leq \ell$. By definition, $\partial_{ii} = \partial_{m_i}$ and $\partial_{i,i+1} = \partial_{m_i} - \partial_{m_{i+1}}$.

Lemma 5.22: For each $u \in F$ and $1 \leq i \leq j \leq \ell$ one has:

$$\partial_{ij}(u) = \sum_{\substack{i \leqslant i_1 \leqslant j_1 \leqslant j, i \leqslant i_2 \leqslant j_2 \leqslant j \\ j_1 - i_1 + j_2 - i_2 = j - i}} c^{j, j_1, j_2}_{i, i_1, i_2} \left[m_{i_1, j_1}, u m^*_{i_2, j_2}\right] \tag{5.14}$$

in the notation (5.11), where the coefficients are translation-invariant integers:

$$c^{j+1, j_1+1, j_2+1}_{i+1, i_1+1, i_2+1} = c^{j, j_1, j_2}_{i, i_1, i_2}.$$

Proof: We proceed by induction on j − i. If j = i, we have $\partial_{ii}(u) = \partial_{mi}(u) = [m_i, um_1^{-1} i]$. Next, assume that j − i > 0. Then, using the translation-invariance of the coefficients in (5.14) for $\partial_{i,j-1}(u)$, we obtain

$$\partial_{ij}(u) = \partial_{i,j-1}(u) - \partial_{i+1,j}(u) = \sum_{\substack{i \leqslant i_1 \leqslant j_1 \leqslant j-1, i \leqslant i_2 < j_2 \leqslant j-1 \\ j_1 - i_1 + j_2 - i_2 = j-1-i}} c_{i,i_1,i_2}^{j-1,j_1,j_2} \delta_{i_1,i_2}^{j_1,j_2},$$

Where

$$\delta_{i_1,i_2}^{j_1,j_2} = [m_{i_1,j_1}, um^*_{i_2,j_2}] - [m_{i_1+1,j_1+1}, um^*_{i_2+1,j_2+1}].$$

Furthermore

$$\begin{aligned}\delta_{i_1,i_2}^{j_1,j_2} &= [m_{i_1,j_1+1}, um^*_{i_2,j_2}] + [m_{i_1+1,j_1+1}, um^*_{i_2,j_2}] - [m_{i_1+1,j_1+1}, um^*_{i_2+1,j_2+1}] \\ &= [m_{i_1,j_1+1}, um^*_{i_2,j_2}] + [m_{i_1+1,j_1+1}, um^*_{i_2,j_2+1}].\end{aligned}$$

This proves the formula (5.14) for ∂_{ij} (u). The lemma is proved.

Now we can finish the proof of Proposition 5.21.

Since $m_{i1,j1} \in I_{j1-i1}(F)$, $m^*_{i_2,j2} \in I_{j2-i2}(F)$, and $[m_{i1,j1}, um^*_{i_2-i_2}]$ belongs to the ideal $I_{k+j1-i1+j2-i2+1}(F)$ for all $u \in I_k(F)$, formula (5.14) guarantees inclusion $\partial_{ij}(Ik(F)) \subset I_{k+j+1-i}(F)$ for all k≥0. This proves Proposition 5.21.

We continue to study "difference operators" in F. Let $m_i, d_i : F \to F$, i = 1,..., ℓ, be linear maps. Denote

$$\partial^{(i,j)} := \Delta^{(j-i)}(m_i d_{i+1} \cdots d_j, m_{i+1} d_{i+2} \cdots d_j, \ldots, m_{j-1} d_j, m_j) \qquad (5.15)$$

for all $1 \leq i \leq j \leq \ell$.

For instance, $\partial^{(i,i)} = m_i$, $\partial^{(i,i+1)} = m_i d_{i+1} - m_{i+1}$, and

$$\partial^{(i,i+2)} = m_i d_{i+1} d_{i+2} - 2m_{i+1} d_{i+2} + m_{i+2}.$$

Lemma 5.23: Let $D = (f_1,...,f_n) \in (F^{\times})^n$. Set $m_i = f_i f_{i+1}^{-1}$. Then for each $u \in F$ one has

$$\Delta^{(j-i)}\left(f_i u f_{i+1}^{-1}, \ldots, f_j u f_{j+1}^{-1}\right) = \partial^{(i,j)}(u'), \tag{5.16}$$

where we abbreviated $u'=f_j f_j^{-1}$ and $d_i := \partial_{m_i} + 1$.

Proof: Indeed, for $i \le k \le j$ one has

$$\begin{aligned} f_k u f_{k+1}^{-1} &= m_k m_{k+1} \cdots m_j f_{j+1} u f_{j+1}^{-1} (m_{k+1} \cdots m_j)^{-1} = m_k (\partial_{m_{k+1} \cdots m_j} + 1)(u') \\ &= m_k (\partial_{m_k} + 1) \cdots (\partial_{m_j} + 1)(u') = m_k d_{k+1} \cdots d_j (u'). \end{aligned}$$

Substituting so computed $f_k u f_{k+1}^{-1}$ into (5.3), we obtain (5.16).

Therefore, all we need to finish the proof of Theorem 5.19 is to prove the following result.

Proposition 5.24: Let $m_1, m_2, ..., m$ be elements of F and $\ell \ge 0$ such that, in the notation (5.9), one has $m_{ij} \in I_{j-i}(F)$ for all $1 \le i \le j \le \ell$. Then

$$\partial^{(i,j)} \in End^{(j-i)}(\mathcal{F})$$

for all $1 \le i \le j \le \ell$, where again we abbreviated $\partial_i := \partial_{m_i}$.

Proof: We need the following notations. Let M_I be the linear map $F \to F$ given by:

$$M_I = m_{i_0,i_1} \partial_{i_1+1,i_2} \partial_{i_2+1,i_3} \cdots \partial_{i_{k-1}+1,i_k},$$

Where

$$m_{i',j'} := \Delta^{(j'-i')}(m_{i'}, \ldots, m_{j'}), \qquad \partial_{i',j'} = \begin{cases} d_{i'} - 1 & \text{if } i' = j', \\ \Delta^{(j'-i')}(d_{i'}, \ldots, d_{j'}) & \text{if } i' < j' \end{cases}$$

for all $1 \le i \le j \le \ell$.

Lemma 5.25: In the notation (5.15) one has

$$\partial^{(i,j)} = \sum_I c_I M_I, \tag{5.17}$$

where the summation is taken over all subsets $I = \{i_0 < i_1 < i_2 \cdots < i_k\}$ of $\{1,\ldots,\ell\}$ such that $i_0 = i$, $i_k = j$, and the coefficients $c_I \in \mathbb{Z}$ are translation-invariant:

$$c_{I+1} = c_I,$$

where for any subset $I = \{i_0, \ldots, i_k\} \subset \{1, \ldots, \_\}$, we abbreviate $I + 1 = \{i_0 + 1, \ldots, i_k + 1\} \subset \{2, \ldots, \ell + 1\}$.

Proof. We proceed by induction on $j - i$. The basis of the induction when $j = i$ is obvious because $\partial^{(i,i)} = M_I = m_i$ for all i, where $I = \{i\}$. Note that

$$\partial^{(i,j)} = \partial^{(i,j-1)} d_j - \partial^{(i+1,j)} = \partial^{(i,j-1)} \partial_{j,j} + \left(\partial^{(i,j-1)} - \partial^{(i+1,j)}\right)$$

for all i, j . Therefore, we have by the inductive hypothesis and the translation-invariance of the coefficients c_I :

$$\partial^{(i,j)} = \sum_I c_I M_{I \sqcup \{j\}} + \sum_I c_I (M_I - M_{I+1}),$$

where the summations are over all subsets $I = \{i_0 < i_1 < \cdots < i_k\}$ of $\{1, \ldots, \ell\}$ such that $i_0 = i$, $i_k = j - 1$ (we have used the fact that $M_I \partial_{j,j} = M_I\{j\}$). It is easy to see that for any $I = \{i_0 < i_1 < \cdots < i_k\}$ one has

$$M_I - M_{I+1} = M_{I_1} + M_{I_2} + \cdots + M_{I_k},$$

where $I_j = \{i_0 < i_1 < i_2 < \cdots < i_{j-1} < i_j + 1 < \cdots < i_k + 1\}$ for $j = 1, 2, \ldots, k$. Therefore,

$$\partial^{(i,j)} = \sum_I c_I M_{I \sqcup \{j\}} + \sum_{I,j} c_I M_{I_j},$$

i.e., $\partial^{(i,j)}$ is of the form (5.17). The lemma is proved.

Now we can finish the proof Proposition 5.24.

Recall from (5.4) that $m_{ij} = \Delta^{(j-i)}(m_i, \ldots, m_j) \in I_{j-i}(F)$ (hence the map $u \mapsto m_{ij}u$ belongs to $End^{(j-i)}(F)$) and from Proposition 5.21 that $\partial_{ij} = \Delta^{(j-i)}(\partial_{mi}, \ldots, \partial_{mj}) \in End^{(j+1-i)}(F)$ for all $1 \le i \le j \le \ell$. This implies that for $I = \{i_0 < i_1 < i_2 < \cdots < i_k\}$ one has:

$$M_I \in End^{(i_1-i_0)}(\mathcal{F}) \circ End^{(i_2-i_1)}(\mathcal{F}) \circ \cdots \circ End^{(i_k-i_{k-1})}(\mathcal{F}) \subset End^{(i_k-i_0)}(\mathcal{F}).$$

Therefore, Lemma 5.25 guarantees that $\partial^{(i,j)} \in End^{(j-i)}(F)$ for all $1 \le i \le j \le \ell$. Proposition 5.24 is proved.

We are ready now to finish the proof of Theorem 5.19.

It is enough to apply Proposition 5.24 to the identity (5.16) from Lemma 5.23.

Finally, we are able to finish the proof of Theorem 5.12.

Note that, according to Theorem 5.19, the inclusions (5.4) imply the inclusions (5.5). Lemma 5.16 and Proposition 5.17 show that (5.5) and (5.6) are equivalent (one can see it by replacing fi by f_{n-i-1}, i.e., passing from m_i to m_{n-1}^{-1} for all i). Now the proof follows from Proposition 5.21 Theorem 5.12 is proved.

We will finish the section with a natural (yet conjectural) generalization of Theorem 5.12.

Conjecture 5.26. Let $g = sl_2()$ and $A = A_n$ be as in Theorem 5.12, and F be an object of Alg_1. Then a matrix $g \in GL_n(F)$ belongs to $G_{g,A}(F)$ if and only if

$$g \cdot \mathfrak{g} \cdot g^{-1} \subset (\mathfrak{g}, A)(\mathcal{F}). \tag{5.18}$$

Remark 5.27. More generally, we would expect that for any perfect pair (g,A) an element

g ∈ (F A)$^\times$ belongs to $G_{g,A}$(F) if and only if (5.18) holds.

ACKNOWLEDGMENTS

The authors would like to thank M. Kapranov for very useful discussions and encouragements during the preparation of the manuscript and C. Reutenauer for explaining an important Jacobi type identity for commutators. The authors are grateful to Max-Planck-Institut für Mathematik for its hospitality and generous support during the essential stage of the work.

REFERENCES

1. E. Artin, Geometric Algebra, Interscience Publishers, Inc., New York/London, 1957.
2. A. Berenstein, V. Retakh, Noncommutative double Bruhat cells and their factorizations, Int.Math. Res. Not. 2005 (8) (2005) 477–516.
3. A. Berenstein, V. Retakh, Noncommutative loops over Lie algebras, MPIM Preprint, 2006, No. 131.
4. J. Dieudonné, La géométrie des groupes classiques, 3rd edition, Springer-Verlag, Berlin/Heidelberg/New York, 1971.
5. B. Feigin, B. Shoikhet, On [A,A]/[A,[A,A]] and on aWn-action on the consecutive commutators of free associative algebra, Math. Res. Lett. 14 (5) (2007) 781–795.
6. A.J. Hahn, O.T. O'Meara, The Classical Groups and K-Theory, Springer-Verlag, Berlin, 1989.
7. M. Kapranov, Noncommutative geometry based on commutator expansions, J. Reine Angew. Math. 505 (1998) 73–118.
8. M. Kontsevich, A. Rosenberg, Noncommutative smooth spaces, in: The Gelfand Mathematical Seminars 1996–1999, Birkhäuser Boston, Boston, MA, 2000, pp. 85–108.

CITATION

1. Arkady Berenstein, Vladimir Retakh, Lie algebras and Lie groups over noncommutative rings, Advances in Mathematics, Volume 218, Issue 6, 20 August 2008, Pages 1723-1758, ISSN 0001-8708, http://dx.doi.org/10.1016/j.aim.2008.03.003.

Index